The Rhythm of Everything

A Journey Through Nature, Science, and Faith

Bonnie Truax

ARMCHAIR
BOOKS

Do we merely want to survive—as the fittest?
Or do we seek to thrive—embracing a life fully lived?
True abundance is found in connection.

Armchair Books LLC
Peoria, Arizona
ARMCHAIR   Send feedback to feedback@armchairbooks.net
BOOKS

Published in the United States by Armchair Books LLC, Peoria

Library of Congress Control Number: 2025904207

Paperback ISBN: 978-1-967099-00-9

EBook ISBN: 978-0-9976522-3-9

Cover designed by Tina Stephens. Images courtesy of iStock by Larysa Pashkevich (Photo ID: 1307188004), Yangphoto (Photo ID: 471657469), 1xpert (Photo ID: 1492870105), and Chinnapong (Photo ID: 961077862). Used under license.

Grammar Rules Broken with Purpose: In The Rhythm of Everything, the capitalization of "Create" signifies the initiation of something entirely new—bringing forth from nothing. In contrast, "create" with a lowercase "c" refers to assembling or combining existing materials, like making art or constructing a bridge from pre-existing elements.

Dedication

I DEDICATE THIS BOOK to my husband, Trinity Montero, for being the most incredible partner I could ever imagine. His patience and unwavering support allowed me the freedom to immerse myself in writing.

To my friends worldwide who have expressed their desire for freedom, I hope my words will serve as a catalyst for you to find personal freedom. May at least one insight from this book be a seed to help you on your journey.

Contents

A Guide to the Journey Ahead

IN 2016, MY HUSBAND and I made a significant life change. We resigned from our corporate jobs, sold our home and its contents. Then, with just a backpack each, we set off to travel the world. While many stories of our adventure are included, this is not a story about me. Rather, it is an exploration into the patterns of nature and the wonders of scientific discovery.

Science has been my lifelong passion, and in this book, I explore a range of scientific findings and theories. Each theory is supported by observation and patterns that point to the truth they unveil.

Along the way, I confront prevailing theories and encourage readers to examine their own beliefs. Ultimately, I share why I believe in God. By the end, I hope you'll be inspired to reflect on the rhythms in your life and the beliefs that shape your view of the world.

This book is divided into three main sections:

- **Parts 1–3:** Explores the intricate patterns woven into nature.

- **Part 4:** Investigates our beliefs about nature and its origins.

- **Parts 5–6:** Examines how misconceptions about our origins contribute to the broken rhythms of our culture and offers pathways to reconnect with nature's design, fostering healing and restoration.

Part 1

The Patterns of Nature

"**Thousands of tired, nerve-shaken, over-civilized people are beginning to find out that going to the mountains is going home; that wildness is a necessity.**" ~ John Muir

The natural world is full of patterns that illustrate the path to vibrant health. Environmentalists study and copy these patterns to help restore ecosystems. Spending time in nature's rhythms can also help heal our bodies.

Every living thing has an important role in keeping the balance of a healthy, thriving environment.

John Muir, *Our National Parks* (Houghton Mifflin and Company), chapter 1.

CHAPTER ONE

Rhythms of the Jungle

"WHO?" MY BOSS YELLED over the phone. "Who, who, who, who?" He continued his rant, demanding I tell him which of our managers made the forecasting error he now had to explain. I had already told him twice I didn't know. Answering again seemed pointless, and I refused to lie. As his repetitive, single-word demand escalated, I sat back and folded my arms. Despite his yelling, I didn't take it as a personal attack. He treated everyone this way. Our team had grown tired of his angry outbursts wasting precious time in our meetings.

A message from a co-worker popped up on my screen: "Has he gone mad?" Another co-worker pinged me, saying her daughter just asked if an owl was on our call.

While I loved IT finance and many of the people I worked with, the increasingly volatile politics within the organization, along with toxic management, gradually drained my loyalty to the company. This marked the beginning of the end of corporate life for me. The constant anger from my boss was both irritating and counterproductive, but the real issue was the breakdown of connection among colleagues. The management team, once united and collaborative, had shifted focus. Leaders had become more interested in building their personal empires than in achieving something meaningful for our customers. This shift had eroded the culture. The joy I once found in work was disappearing, replaced by frustration as our boss hurled angry words across conference tables. Where long hours had once been filled with camaraderie, they now left me craving the connection we had lost. The rhythms of work and life had fallen out of sync.

There was never enough time for the things on my to-do list both at work and at home. Projects with little margin in their deadlines had to keep moving forward despite the need for rewritten code and another release of the application. Late nights and early mornings pushed sleep deprivation well into exhaustion. The continuous stream of back-to-back meetings and conference calls was stretching us all too thin.

Quietly, my husband Trin and I began planning to change the pattern of our lives. Gathering a few of our closest friends, we announced our news: we were selling everything and embarking on a journey around the world. Both of us gave a four-months' notice at work, then we put our home on the market, setting the wheels in motion for our adventure.

A year after leaving the corporate world, I reminisced about the long days and short nights, realizing then how many natural rhythms I had disrupted—sleep was only one of them.

A Swim in the Amazon River

Two years after escaping the corporate rat race, we found ourselves in Manaus, Brazil, in the heart of the Amazon Rainforest.

The afternoon was sweltering hot when Trin and I hopped into a van that transported us to the shore of the Amazon River. There we joined two other couples. Together, we embarked on a series of boats, each one smaller than the last, which led us deeper into the rainforest.

The last boat, crafted from tin, offered just enough seating for our local guide, named Cacà, and our party of six. We sat just above the waterline. Sweat dripped from our faces, soaking our clothing in the stifling humidity of the Amazon jungle.

Cacà navigated our small boat through the river system and eventually pulled up to a rickety dock. A large, enclosed gazebo, visible from the shore, would become a gathering place for meals and evenings spent with our new friends.

We carefully disembarked from the boat onto the dock and explored what was to be our new home while we stayed on Cacà's land. Grabbing our packs, we walked down a path extending into the jungle to find a cluster of huts crudely assembled. Gaps between their graying planks left us never fully separated from the jungle.

Over the next several days, Cacà shared his way of life with us, even mentioning that he had married his cousin. He was a short, stocky man, built for the demands of the jungle. This was his birthplace, where he had grown up fishing, hunting, and mastering the rhythms of the wild. This was Cacà's life. He had never ventured beyond Manaus, the remote city we had just come from, 70 miles (113 kilometers) away as the crow flies. The journey to the city covers more than twice that distance due to the lack of roads; most travel is done by navigating the waterways.

"There's nothing to do in the city, not like out here," Cacà said in accented English, his words only slightly misaligned. He spoke slowly and deliberately, his cadence matching the gentle way he bathed his child. With each careful scoop from the bucket—water he had drawn from the river—he poured a steady stream over the little one, the unhurried flow mirroring our conversation.

"This is a great place to raise a family," he added, glancing up with a quiet sense of contentment.

Just a few days prior, Cacà showed our group how to chum the water around our little tin boat to attract piranhas. Then he gave us raw chicken and hooks to catch these voracious fish lurking beneath us. Piranhas quickly attacked our lines, their hunger unyielding even when hoisted out of the water. One piranha took a bite out of Jeremy's hand, one of the six in our fishing group. We caught enough piranhas to fill our appetites at dinner that evening. Jeremy and his wife, Alia, had the extra treat of a tale to tell when they returned home to Washington, DC.

Everything in the river appeared to possess formidable teeth. Creatures in the water could bite (several fish), squeeze (anacondas), or sting (electric eels) the life out of their prey. Once, we nervously watched as Cacà, who had been fishing behind us in the boat, reached into the Amazon's dark waters to retrieve his catch. There was our host and navigator, shoulder-deep in the dangerous river, grappling with something that was fighting back. The visibility into the water extended only a few inches. Tannins from jungle leaves infuse the river, giving it a dark color as if someone steeped tea too long. We couldn't discern what Cacà was battling until he emerged with a huge smile and an even larger fish.

"Dinner!" he declared.

After our meal, we embarked on an evening ride in search of caimans, which are related to alligators. The jungle was at once beside and in the river, with no clear separation. Trees crowded around us as we navigated through them in our small dinghy. Our flashlights scanned the water's surface, revealing the caimans' glowing eyes. Suddenly, Cacá reached into the dark water and grabbed something. He stood with a small caiman in his hand, then passed it around so we could rub its soft belly and examine the rugged camouflage scales on its back.

"I recommend you no touch anything," Cacá cautioned the next day before docking on another shore and leading us into the jungle. Using his machete to clear the path, he explained scorpions and other dangerous bugs easily hide on the trees.

"Did you play here as a kid?" Alia inquired.

"No," he responded with an alarmed look, "it is too dangerous for a child."

Single file, the six of us followed our guide, weaving through the dense vegetation.

Cacá demonstrated how to craft a trap from vines. Then he grabbed a babassu fruit, sliced it open with one slash of his machete, and pulled out something resembling a large grub. It was the larva of the beetle Pachymerus nucleorum. After finding a few more large, white, squirming larvae, he handed them to us, making the gesture of hand-to-mouth, indicating we should eat them. These larvae, having spent their entire life to this point in these small, elongated, coconut plants, were surprisingly tasty. The coconut flavor was distinct, with only a slight aftertaste of "guts."

Using the dull side of the machete, Cacá struck the gill-like extension of the lupuna tree. The tree, seemingly impervious to any damage, echoed the strike through the jungle.

"This is the telephone tree," he announced. "While hunting, we use this as a call for assistance or to indicate a catch. If we hit it ten times, it means we are lost, and someone will come to find us."

The sweltering heat and the hike through the jungle prompted us to consider Cacá's invitation to swim in the river.

"It's what the locals do," Cacà encouraged. Despite the limited visibility in the water and the awareness of potential dangers, the water was pleasant. At that point, swimming in the Amazon River somehow felt normal. Cacà had obviously survived many dips himself, even when fighting with something in its depths.

After all, it was a "good place to raise a family." In remote places like this, far from cities, I recognize and experience the patterns of the earth, how they flow together in perfect harmony. It didn't take long for us to fall into the rhythm of the rainforest. Each night, we were enveloped by a darkness that erased all visual distractions while the sounds of jungle life gently lulled us to sleep. The soft pastels of the rising sun, along with the chorus of birdsong, stirred us awake when morning arrived. Here, life moves at its own pace—the river flows steadily, leaves reach for the sky, and wildlife patiently awaits its next meal, knowing it will surely come. Everything is interconnected, often so tightly that the lines of separation blur. This seamless blend of life filled my heart with a deep sense of peace and wonder. There was no rush—no blaring alarm clock jolting us into the day, no machinery drowning out the natural world, not even the hum of electricity to distance us from the calming sounds of nature.

Throughout our extensive journey across South America, Trin and I often lost track of the conventional markers of time—days, weeks, and even months. The arbitrary naming of each moment, with a date and time, lost significance while living off-grid.

In these realms, far from civilization—if the relentless pursuit of *more* can even be called civil—I found unparalleled peace. From this vantage point, what we left behind in our high-paced corporate jobs seemed neither restrained nor refined. Even the most exquisite dining in New York City only offered a fleeting glimpse of true luxury. True luxury is not measured by wealth of gold. It is found in the jungles of Brazil, the outback of Australia, the woods of Sweden, the dramatic peaks of Patagonia, the pristine shores of South Georgia in Antarctic waters, and the breathtaking red rock landscapes of Utah. In these off-grid locations worldwide, the grandeur of life's patterns shines brighter than gold.

We left the alarm clock behind, yet it took a full year without that horrid sound each morning to regulate my sleep pattern. Once I left the corporate world, I began resting at sunset and waking up naturally. Allowing my body to follow its natural circadian rhythm liberated me from chronic fatigue and over-caffeinated days—a small but transformative step toward freedom.

The simple luxury of getting enough sleep initiated a change within me. Having the time merely to exist, free from the pressure of making every moment "productive," opened up my ability to discern patterns all around me. It granted me the time to process life.

What was it about this place, this immersion in the natural world, that brought such profound peace and a sense of rightness to my sleep, to my very being? Part of the answer lies in the science of our internal clocks: the circadian rhythm.

Circadian Rhythms

The rotation of our planet and its orbit around the sun resembles a graceful dance. With a subtle wobble back and forth, it orchestrates the rhythmic cadence of days, nights, and seasons. It inextricably influences all life on Earth.

It's called the circadian rhythm in science: a cyclic pattern lasting 24 hours (to be exact, 23 hours and 56 minutes) that our bodies follow. Cues such as light and darkness, along with various environmental factors, signal our brain to release hormones. These hormones play a pivotal role in regulating sleep, body temperature, and metabolism. This cycle can vary from one person to the next.

> Humans are not sleeping the way nature intended. The number of sleep bouts, the duration of sleep, and when sleep occurs has all been comprehensively distorted by modernity.[1] ~Matthew Walker, neuroscientist and sleep expert

The circadian rhythm is just one of the many patterns that contribute to greater health and peace in our lives. Throughout recorded history, civilizations have acknowledged the changing seasons, employing various ornate methods to track them. Seasons have served as guides for planting and harvesting as well as signals for migration. Many such patterns are evident and beneficial for life, even before we discovered the intricacies driving each cycle. We don't need to comprehend why everything in nature is as it is before we follow the obvious patterns surrounding us.

Personally, aligning my life with the cycle of the sun and finally getting enough sleep was a major step in restoring my natural rhythm. Finally, I could go through a day without exhaustion. My sense of peace grew, and instead of dozing off whenever I sat down, I could fully engage with life, processing it with clarity. It was then I began to see how everything had a connected rhythm.

Our universe unfolds with clues to understanding how these rhythms collaborate and shape life. The Earth's rotation governs the circadian rhythms essential to our bodies. Galaxies move in a grand cosmic dance, telling a story far greater than a well-rested night. These patterns don't just sustain life; they elevate it beyond mere survival—they are exquisite.

CHAPTER TWO

Mysteries of the Daintree

THE DAINTREE RAINFOREST IS often called mystical and otherworldly. Trin and I were excited to visit Far North Queensland, Australia, where this amazing natural wonder is located. Maybe its appeal comes from the trees that resemble those in Jurassic Park or from the cassowary, the world's second heaviest and most dangerous bird. It could also be the plants that entangle hikers, forcing them to pause until they manage to free themselves. Personally, what fascinated me was the abundance of unique life. It surpassed our expectations and captured my imagination.

From the Mount Alexandra lookout, fog settled just below the canopy, giving giant ferns the appearance of floating. Walking through the dense undergrowth, surrounded by ancient trees reaching high into the sky, I grasped the inspiration this jungle could provide for many mysterious and wonderful things. The trails felt like pathways to the past in this oldest, continually-surviving rainforest.

The Daintree is not only visually stunning; it also sounds as if Jurassic Park borrowed its soundtrack from this place. I stop and look around, trying to identify the source of each new tune. Camouflage is perfected here, often concealing the singer of a song. To learn which creature is calling, I must sit still and wait.

Rain filters through the canopy, soaking my clothes and cooling me as I quietly observe, waiting for the forest to reveal its secrets. The cacophony of cries intensifies, each chorus further enveloping me in the vibrant world around me. Movement above catches my eye—a tiny beak twitches in time with the unfamiliar sounds. The sonorous, intricate tunes make me question my eyes. How can this miniature creature fill the jungle with such a captivating melody?

Then the forest quiets, and the soft gurgling of a stream gently flowing takes the stage. Water follows paths through the understory, its purity giving visibility to every detail of the stream bed.

Shades of green surround me, crowding together and appearing so thick as to be impenetrable. Vines twist and reach out, so entwined that it is hard to discern where one plant ends and the next begins. Leaves of diverse shapes and sizes adorn the jungle. Colorful fruits grow from the branches and trunks while orchids sprout from the crevices of trees.

I glance down the trail and notice Trin untangling from a wait-a-while vine. These vines seem to grow along footpaths where disturbed ground gives them room to expand. Their tiny barbs hook onto anything passing by. Movement often entangles the trespasser even more into the vine. It takes a while to disentangle all the hooks from clothing and skin.

The wait-a-while vine may be a nuisance, but it's nothing to worry about compared to the stinging tree whose minuscule hairs inject a potent neurotoxin, causing painful sensations. Pain persists for hours to days, landing some in the hospital for a week, with recurring episodes of pain that occasionally last for years. Thankfully, we did not come into contact with this vine.

Destructive cyclones come and go, but the Daintree recovers. Young saplings await their time. When the canopy above breaks, they race to the top in a competition for sunlight, quickly filling in the gap.

The rhythm is remarkable. Continents have shifted, earthquakes have toppled mountains, and the world constantly changes, but this rainforest holds on.

As we walk, we keep our eyes out for the elusive cassowary. We want to observe this magnificent bird here in its natural territory.

The cassowary sports a casque on its head resembling a large horn-like crest. This crown may look deadly, but many believe it is used to release heat, something quite useful when temperatures climb and humidity in this region of Australia is high. The cassowary's weapon is the dagger-like talons on its large feet that can deliver a kick capable of inflicting fatal injuries to humans.

They can be aggressive, especially when protecting their young, which is the job of the male cassowary. The female lays the eggs but then goes on her way, leaving the male to hatch and raise the young. Many of the ways of the rare and endangered cassowary are still a mystery to scientists.

Warning signs at trailheads provide instructions for cassowary encounters: avoid running, back away slowly, and use a tree or other object as a barrier.

As we searched the trail for a cassowary, the grunts of a feral pig put us on alert. Soon, we came upon a tree bearing fruit directly on its trunk—easily within reach, unlike most fruits that dangle high in the canopy.

This is called the Ryparosa kurrangil, and it is endemic to Australia. The Daintree is the exclusive habitat for this plant, spanning from Mossman to Cape Tribulation. This fruit grows low enough for the flightless cassowary to reach and feast on.

The Ryparosa kurrangil is an endangered plant whose seeds only have a 4% chance of germinating when they fall to the ground. However, passage through the gut of a cassowary increases the germination rate to 92%. Scientists have been unable to replicate this process in the lab.

Ryparosa kurrangil's dependence on the cassowary is just one small piece of the Daintree lifecycle. It is an elegant example of the codependencies among so many living things.

In the Daintree, there is also a large, bright blue plum fruit with sap that is toxic to most animals and humans. Most species avoid this fruit; therefore, the cassowary, which is immune to the toxin, has an abundance available. Interestingly, this fruit's germination also relies on the cassowary's stomach acid and digestive process to thrive.

So two plants depend on the cassowary. The cassowary, who consumes large amounts of fruit, in turn depends on the abundance of these plants. The Daintree ecosystem is a profoundly interconnected web of life. Even fungi play a pivotal role in the cycle, transforming death into nutrients for new life.

The wild boar did not appear on the trail that day, nor did the colorful cassowary. Thankfully, we had the opportunity to observe a few of these magnificent birds during our time in Far North Queensland. Often, when spotted, the cassowary swiftly vanishes into the foliage. The first one we saw along the road did just that, but the next one, upon spotting me, boldly approached, its eyes fixed on me.

I never intended to get that close. Some scientists suggest that when a cassowary is accidentally hit on the road, it's not due to stupidity but rather arrogance. I sensed that arrogance that day. As he circled me, his gaze seemed to ask, *What are you doing on my trail?* I stood still, careful not to alarm him. Who can predict the nuances of nature or why he gracefully left me alone that day?

In diverse landscapes across the globe, I have learned to be still and observe the patterns of life in the wilderness. Time spent mindfully being and learning by observation is a cherished freedom, the impact of which continues long after I have left the forest.

Every piece, from the soil to the weather, is integral to an ecosystem's health. Each organism, throughout its life cycle, contributes to the rich nutrients in the soil. It is a rhythm where every species plays a sustaining part. When people clear a portion of the Daintree, that area loses the nutrients necessary to support the forest. Scientists have tried, but to date, they are incapable of replanting any rainforest in a lab. The Daintree cannot even be reforested by human effort in cleared areas.

The Daintree, however, can expand and recolonize itself up to a meter a year, as if it were one organism. It revitalizes the soil with lost nutrients gradually. It is an ecosystem that cannot be initiated with only one plant or animal; the entire network is essential for many of the species to exist.

The interwoven and self-sustaining cycles of this rainforest lead some to call it a closed ecosystem.[1] But it is not entirely closed. It would not survive without the rhythms of the sun providing energy for life, the wind strengthening the trees, and the rain bringing nourishment.[2] The storms that bring down old trees contribute to the cycle, turning them into nourishment for new life, which springs up once again to cover the canopy.

Similarities abound on the opposite side of the globe in the Amazon Rainforest.

The Sahara-Amazon Dance: A Transatlantic Ecosystem

Walking through the Amazon jungle is a fully immersive experience. The air hangs heavy with humidity, thick with the scent of decaying leaves and blooming orchids. Everywhere you look, life and death intertwine, creating a sense of one vast, interconnected system. This web of life depends on the constant cycle of growth, decay, and renewal.

The Amazon jungle teems with so much life that it consumes nearly all the oxygen it produces. Like the Daintree, it is not a closed system; instead, it is a key component of a much larger pattern extending across the globe that provides our planet with oxygen.

The Andes Mountains form a wall around the western edge of the Amazon basin. This wall captures moist winds from the Atlantic Ocean, compelling them to cascade down over the basin, where incessant rain over the Amazon jungle creates the world's most biodiverse region. However, it also leaches nutrients from the soil, washing them into the ocean. Huge algae blooms of phytoplankton develop where the fresh, nutrient-rich water pours into the ocean. Phytoplankton consume these nutrients and incorporate carbon into their bodies. When they die, the decomposition process releases some of this carbon as carbon dioxide (CO_2) into the water, which is released into the atmosphere.

As the plume of freshwater extends further into the ocean, diatoms and cyanobacteria become prominent. These organisms, often living in a symbiotic relationship, use nitrogen from the air and, importantly, carbon from the water to fuel their growth. Through photosynthesis, they produce a significant portion of the oxygen on our planet, estimated to be between 20% to 50%. The diatoms eventually die, their silica shells and other organic matter settling on the ocean floor.

On the other side of the globe, vast quantities of cyanobacteria thrive within the biological soil crusts (also known as biocrust) of the Sahara Desert.[3] When these organisms die, their remains, along with other components of the biocrust, contribute to the dust in the dunes and dry lake beds of the world's largest non-polar desert.

Strong winds lift the dry, nutrient-rich dust from the Sahara, including the remnants of the biocrust, creating massive, tan-colored clouds that blow across the Atlantic Ocean. When these dust clouds reach the Amazon basin, rainfall and other deposition processes bring the dust particles, including the cyanobacteria remains, down to the rainforest canopy and eventually into the soil, replenishing it with nutrients.[4]

Diatoms and cyanobacteria produce oxygen essential for life.

Remains of cyanobacteria contribute to the nutrient composition of the Saharan dust. The Saharan dust provides vital nutrients to the Amazon basin. The Amazon basin, through its river's outflow, delivers essential nutrients that help sustain populations of diatoms and cyanobacteria in the ocean. Our globe is a symphony, with the sun's heat driving the wind that plays the strings of this interconnected orchestra.

The Sahara Desert is still not fully understood, and there are ongoing debates about its formation and composition. Yet, the interconnected patterns we are just beginning to decipher leave me in awe. It is a story of life, death, and connection. It is not just the Daintree and Amazon basin; it is our entire globe and universe that are interdependent.

Our world and universe sing a song of exquisite and intricate dependencies that work together to sustain life. The earth has much to teach us.

Speak to the earth, and it will teach you. ~ Job 12:8, NIV

We could endlessly marvel at the symmetry and patterns that have been woven into the world. The more we discover, the more we see them intertwine.

Nature has a cycle that sustains life. Every species, including humanity, plays a specific role, and at the end of life, each gives nutrients that sustain the pattern. Death, the ultimate sacrifice, brings life. It is the basic pattern of our world, a symmetry played repeatedly, underscoring its importance.

CHAPTER THREE

Life Beneath Our Feet

NATURE'S PATTERNS EXTEND BEYOND circadian rhythms driven by celestial movements, the interplay of forests and deserts, and the winds that flow between them: they also reach deep into the soils beneath our feet.

The Mighty Ant

In the northeast corner of Costa Rica, near the Nicaraguan border, lies the laid-back town of Tortuguero. There are no cars in town, and no roads to get there. We arrived on a riverboat from the nearest town, thirty kilometers away. Our bungalow was a simple structure of corrugated aluminum, arching in a U-shape over a cement floor. Bill, the owner, had built several others, but the relentless rainforest kept consuming them.

"The ants here are kings; they carried away my last house," Bill informed us. He initially attempted to construct a treehouse using treated telephone poles for support, but the ants swiftly ate through the poles and eventually carted them away one grain at a time. Ants here can even infiltrate concrete. Nothing is safe from the ants.

Not far from our bungalow, a highway of leaf-cutter ants was bustling. Hundreds of perfectly cut pieces of green leaves formed a procession along a well-trodden path from tree to nest. Each ant, bobbing back and forth, carried a piece as if presenting a five-star dinner. Intriguingly, leaf-cutter ants don't consume leaves; they transport the cut pieces back to their nest to fertilize their farm. These ants cultivate Lepiotaceae fungi to feed their larvae.

Ants were creatures that increasingly astounded me during our travels—so small yet so mighty and amazing.

While in Nicaragua, we encountered a man who had established a bug museum in León. Having dedicated his life to studying bugs, he eagerly shared insights into the lives of each insect in his collection.

He spoke of the cleaner ants (army ants) that don't build nests but function like roaming nomads. Coastal residents willingly let these cleaner ants into their homes, allowing them to eliminate all vermin, including other ants and cockroaches. When the cleaner ants enter a home, it is a great excuse for the human occupants to spend a day at the beach. The ants efficiently complete their task by nightfall, leaving the family with a bug-free home. It proves to be a healthier and more effective alternative to insecticides. For the frugal among us, it is also more cost-efficient.

In Australia, we repeatedly found leaves sewn together, like baskets, on the ground. We discovered that these were the abandoned nests of weaver ants. Despite their small size, these ants possess a beautiful, translucent green abdomen that, surprisingly, tastes like lime. They gather green leaves and adeptly stitch them together with silk, suspending these nests in trees. Fiercely territorial, they not only defend their nests from predators but play a crucial role in protecting the surrounding trees, preying on harmful insects and promoting the health of forests and farms.

The diversity of ants and their roles on Earth is astounding. Each species plays a critical role in environmental health—even fire ants, which help control populations of termites, roaches, and other pests. Just don't step on their nests! I tried that once as a kid. It was painful, and no, it wasn't one of my many experiments that I was always getting into trouble for.

Research estimates there are 20 quadrillion ants on Earth (that's 20 followed by 15 zeros), with their biomass surpassing the combined biomass of wild birds and mammals.[1] Ants contribute to soil aeration, nutrient replenishment, seed dispersal, and invasive species control. Each colony plays a unique role in the interconnected rhythms of nature.

> Ants certainly play a very central role in almost every terrestrial ecosystem.[2] ~ Patrick Schultheiss, entomologist

> Go to the ant, you sluggard; consider its ways and be wise! It has no commander, no overseer or ruler, yet it stores its provisions in summer and gathers its food at harvest. ~ King Solomon, 970–931 BCE[3]

Ants are tiny, mighty, and numerous. Yet, even more delicate than these creatures is a kingdom whose biomass on Earth is exponentially greater than that of all humans combined: the mysterious and wonderful world of fungi. Scientists estimate that a million more fungal species remain to be discovered, highlighting how much we are still learning about this essential life form.

The Intelligence of Slime Molds

Fungi do not have central processors or a brain, and they do not possess awareness or intelligence as we understand it. Yet, they exhibit communication speeds that surpass our current understanding. The term "intelligence" will be used loosely in this book to describe the communication within a mycorrhiza network, because we lack the terminology to describe how they operate with such efficiency and speed. It is essential to acknowledge that fungi operate without anthropomorphic traits, but we will use these terms for a lack of better descriptors.

In experimentation, slime molds[4] "figured out" the most efficient rail paths through Tokyo in just twenty-six hours, something which took engineers hours of research, experience, education, and planning.

Atsushi Tero from Hokkaido University conducted an experiment using the slime mold Physarum polycephalum in a petri dish. Tero replicated a scaled-down version of Tokyo and its rail-connected cities by situating oat flakes where cities with railway connections were located and using light to represent the mountains. Then he placed the slime mold where the city of Tokyo would be. Mold spreads out indiscriminately but inherently seeks food (oat flakes) and avoids light (mountains, in this model). Upon encountering food sources, specific lines of the mold strengthened, while lines without food disappeared. Remarkably, within twenty-six hours, the slime mold created a pattern resembling Japan's current railway system, demonstrating its efficiency in creating the most effective route between two points.[5] The slime mold outperformed human capabilities in transportation network planning.

Wolfgang Marwan of Otto von Guericke University highlighted the significance of Teros's work, referring to the mold as a "complex biological phenomenon." Marwan said, "You see they optimize themselves somehow, but how do you describe that?"[6] Yet, without a central processor fungi is a critical component of the forest.

Next time someone calls you a slimy dirtbag, take it as a compliment—faster and more effective than scientists, and a veritable jungle of wonder. But I don't suggest using this term for others, as they might not understand.

The Networks of the Forest

While we spent a summer in Sweden, a neighbor introduced us to Chanterelle mushrooms. We first foraged in the forest just outside our door for this complex, savory fungus. The earthy, almond fragrance guided us to small patches as we explored the mossy undergrowth. After discovering a patch, we promptly marked the GPS location to return after each heavy rain for new delicate, orange miniature wonders. The location of Chanterelle mushroom patches is a closely guarded secret among the Swedish people, who prize them as a culinary delicacy and an important nutrient source. We relished every bite, our enjoyment heightened by my fascination with the role of fungi on our planet.

There are over ten thousand edible mushroom species known to us so far. Mushrooms, the fruiting bodies of fungi, represent only a fraction of the mycelial network in the soil—a network of tiny, delicate hyphae (see Figure 3.1).

In 1926, researchers discovered penicillin in fungi, the first naturally derived antibiotic. Since then, people have been using fungi in various health contexts, including anti-virals. Studies indicate that turkey tail mushrooms can even enhance cancer treatment.[7]

Fungi can also offer significant benefits to bees. Prominent mycologist Paul Stamets has researched the use of *Metarhizium* mushroom extracts to control mites that threaten bee colonies, offering a potential pesticide-free solution. Separately, researchers at Washington State University have developed a strain of *Metarhizium* that can withstand the temperatures found within beehives.[8]

Far beyond health benefits, fungi play a crucial role in our environment and soil. They break down dead trees and vegetation into nutrients, serving as the forest's digestive tract. Fungi can break down any hydrocarbon-based substances, including oil spills.

More than a digestive tract, mycelial networks play a vital role in the intricate relationships within the forest. Serving as a communication highway and nutrient marketplace, they support life, transfer life-sustaining resources among organisms, break down organic matter, and stabilize carbon in the soil. Carbon stabilized by fungi can be retained in the soil for centuries or even millennia.[9] These networks are like a hidden world beneath our feet.

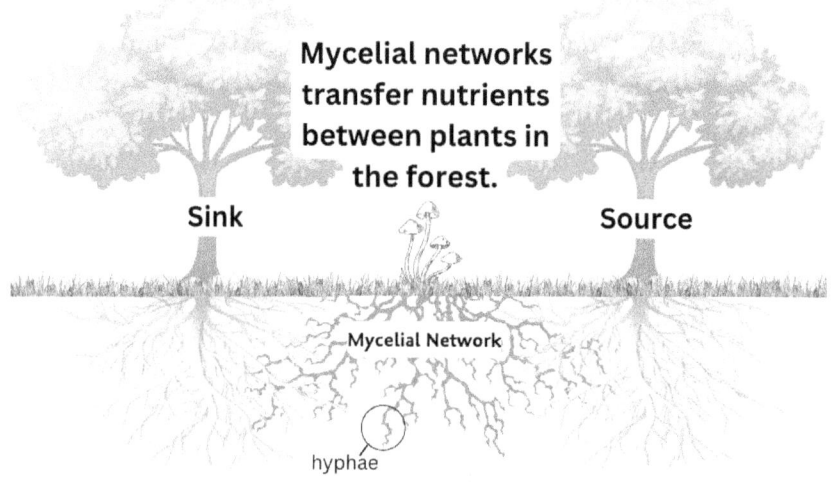

Figure 3.1 Mycelial Networks

In 1885, Albert Bernhard Frank coined the term "mycorrhiza" to describe the symbiotic relationship between fungi and plants. He emphasized fungi's crucial role in the environment, highlighting their partnership with plant roots. These fungi help plants absorb essential elements like water, carbon, nitrogen, and minerals, while the plants, in turn, provide sugars to sustain the fungi. Mycorrhizal networks create an interconnected system among plants and trees, functioning as a source-sink network. In this system, nutrient-rich "source" plants share their surplus through fungal connections, efficiently directing resources to "sink" plants in greater need.[10]

Mycorrhiza networks enable communication and exchange among plants. They allow trees to recognize kin and share essential resources. When a tree is attacked—by pests, disease, or environmental stress—it releases distress signals through these fungal connections. Neighboring trees detect these signals and respond by producing defensive chemicals or hormones, such as tannins to deter herbivores. They can also produce volatile organic compounds to attract predator insects that feed on the attackers. This communication helps trees support and protect one another, reinforcing the mutualistic relationships within the ecosystem. This interconnectedness is key to the health and resilience of the entire forest ecosystem.

The mycelial network's ability to gather carbon efficiently from plants and trees and store the excess carbon in the soil is highly critical. Imagine the significant carbon sequestration potential that would happen if we abandoned the practice of plowing between trees in an orchard and instead fostered a thriving undergrowth. Not only would it sequester more carbon, it would also provide trees with the critical communication networks they require to thrive. I can't help but wonder if the practice of plowing and leaving the earth bare under the scorching sun in our Florida orange groves is contributing to their decline. After years of abuse, the trees, isolated from each other and their own natural defenses and subjected to chemicals, are losing the battle for life.

Scientists estimate the biomass of all fungi on Earth to be twelve gigatons, surpassing the combined biomass of humans.[11] Estimates suggest that only 6% of fungi are cataloged. Each handful of soil is a microscopic jungle. There is so much to discover! Understanding the connections and relationships between fungi and their environment is essential for recognizing patterns and appreciating the symbiotic nature of our ecosystem.

Fungi serve as food, medicine, the digestive tract of the forest, and a communication highway. They function as a defense system and play a crucial role in regulating our planet's carbon levels.

Our ecosystem exhibits a profound symbiosis, emphasizing that each entity contributes to the overall narrative. Organisms separated from this interconnected system typically face challenges or decline. Every life form contributes and is an active participant, essential for fostering a thriving environment. This includes you and me.

In the grand symphony of nature, humans play a role with the unique capacity to choose between acting as a parasite or a vital, thriving part of the network. The rhythms of the natural world offer profound insights into our place within it.

CHAPTER FOUR

Restoring Natural Rhythms

OBSERVING THE NATURAL PATTERNS of our world imparts valuable lessons on caring for our planet. Deviating from these patterns can lead to destructive consequences.

Reversing Desertification

Allan Savory, an ecologist and the co-founder and president of the Savory Institute, dedicated his career to combating desertification, a form of land degradation. In his 2013 Ted Talk, he first recounted his initial failures. Then he shared a pivotal discovery in the patterns of nature that transformed his efforts, enabling him to achieve significant results.[1]

Having immersed himself in the study of ecology for years, Savory engaged in a project that involved culling forty thousand elephants, aiming to rejuvenate dying desert soils. Prevailing scientific theories of the time suggested that the herds' activities and movements were detrimental to the land. But when the elephants were gone, the land experienced further degradation. This pattern of deterioration was not unique to this project. Researchers observed it globally in managed lands. Removing animals resulted in continued deterioration.

Savory's perspective transformed when he observed how grasslands suffer if they entered the dormant season without proper biodegradation. The grazing process turns end-of-season grass into nutrient-rich manure as it passes through the guts of herbivores. The herds' stomping power, rather than destroying the land, broke down the grasses and mixed them with manure. Elephants thus play a crucial role in the biodegradation process. By using livestock to mimic natural herd patterns, Savory witnessed dramatic healing in once-barren lands. The trampled grass mixed with dung created a rich mulch cover, promoting fresh growth and enabling the land to hold moisture. Less runoff also meant less erosion and a reduction in flooding. The land began to heal. Mimicking nature restored the ecosystem to its original, self-sustaining pattern.

This revelation marked a turning point for Savory. Instead of viewing animals as the cause of degradation, he recognized that nature operates through a symbiotic relationship between animals and the environment. Removing components disrupted the delicate balance, further degrading the land. Savory acknowledges the complexity of desertification and land management, emphasizing that we are still learning how to nurture our land.

It is important to observe the designs of nature. We do not need to create something new or improve nature. Observation and experimentation suggest that we would be better off following the patterns already in place. Altering one part of the ecosystem without understanding the entire pattern can lead to unintended consequences. Labeling one aspect bad, like elephants roaming the plains or cattle grazing, disrupts the pattern. Herd management may be necessary, but eliminating one part of a cycle can lead to a domino effect of destruction.

Modern humans have bastardized the ancient idea of dominion, turning it into one of rulership and control. True dominion is not about taking what we want but rather recognizing our role within the patterns and playing our part to nurture, give, and, dare I say, serve.

The only dominion beneficial to nature is one that respects patterns, co-dependencies, and the importance of all life. There is a pattern set in motion that, if followed, leads to better health for soil, plants, animals, and humans. When we break those patterns, we sabotage life.

In our efforts to restore land, we may sometimes get it wrong. Scientific proposals must be fine-tuned or corrected with each discovery, but the suggestions that consistently work are those that mimic nature. Nature's patterns, unchanged for millennia, remain the most effective guide. Circumventing nature to forge our own path repeatedly fails. Humans have not devised a system that functions better than nature itself.

> Not a single long-term tillage system on earth exists without an herbivorous component.[2] ~Joel Salatin

Wolves in Yellowstone

Grazing animals contribute positively to the land, but maintaining healthy land involves more than just grazing and bio-processing to create fertilizer; the pattern of their movement is also crucial.

In the 1930s, wolves were hunted to extinction in Yellowstone National Park. Mysteriously, the beaver population declined. The park once teemed with a large, healthy beaver population, but by 1995, only one beaver colony remained in the park.

The mystery of the declining beaver population lay in the natural cycle that was broken when wolves were eliminated. The absence of wolves left elk populations without predatory pressure, causing herds to expand and linger near the water's edge—an area that was previously dangerous. Unchecked, they consumed willow trees crucial for beavers to build dams and survive winters. Consequently, the beaver population dwindled due to resource competition with elk.[3]

In 1995, wolves were reintroduced to Yellowstone. The predatory pressure prompted elk to move more frequently, restoring their natural migratory pattern. As a result, they spent less time lingering near the streams and eating down the willow trees, leaving more for the beaver colonies. Since the reintroduction of wolves, the beaver colonies have rebounded, with an estimated 108 colonies in 2021.[4]

This Yellowstone example highlights a vital ecological pattern: removing one element disrupts the entire system, triggering downstream impacts that compound over time. The ecological system is intricate. It is a finely-tuned harmony—much more than a mere machine. Ongoing studies in Yellowstone are still capturing additional benefits from the reintroduction of wolves.

The Polyface Farm Model

In 1961, William and Lucille Salatin migrated with their family from Venezuela to the USA and bought a farm that was eroded and abused.[5] The land had depleted soil, eroded hillsides, and exposed bare rock.

The Salatin family transformed a once deemed useless property into the highest-producing farm in the region, but it took years of work. Without the use of chemicals, they restored the soil and now it serves as a reliable carbon sink and an integral part of the farm. At the Salatin Polyface Farm, they produce abundant pesticide-free, chemical-free food. Both plants and animals contribute to the nutrient cycle within their natural rhythm.

Joel Salatin, son of William and Lucille, now manages the farm and practices dominion as a nurturing role, caring for the life on his land. Rather than imposing his will, he works within natural patterns, understanding the importance of observation, patience, and respect for natural processes. Polyface—meaning many faces—incorporates various animals, including chickens, pigs, turkeys, rabbits, and cattle, allowing each to express their uniqueness by doing what they love. Pigs get to wallow, chickens get to scratch, and cows chew their cud as they gaze out over the verdant hills, contemplating life—at least that is what they look like they are doing. Each animal contributes to the farm by fulfilling its natural role.

By recognizing and using natural patterns, Salatin has optimized green space usage, turning these spaces into effective carbon absorbers. Cows play a crucial role in the fertilization of these spaces, eliminating the need for harmful agents. Despite the absence of toxic chemicals, pathogen levels on the farm are significantly lower than on industrial monoculture farms that sever natural patterns and treat animals as mere machines.

Inspired by herd predator patterns, Salatin mimics migratory movements by keeping cows together as a herd and moving them frequently. This prevents overgrazing and ensures a more even distribution of the manure that they contribute across the fields. Chickens follow the cows, fulfilling their role of eliminating bugs by scratching through cow patties and enjoying their favorite meals.

Healthy grass produces more oxygen per square foot than trees covering the same area. Grass is also a much more reliable carbon sink than trees when considering forest fires.

Salatin's nurturing approach extends to winter, when cows are in the barn. A layer of wood chips generated from forestry work, referred to by Joel as a "carbonaceous diaper," provides comfort for the cows and keeps the barn smelling fresh. Whole shelled corn is occasionally mixed in as the carbon-nutrient diaper builds. In the spring, pigs complete the job of stirring the compost pile. They are in heaven, rooting and searching for the hidden corn. This produces a healthy compost for the spring fields.

Salatin uses a carbon-manure mix in his barn, a stark contrast to the concrete slabs and liquid manure vats common in conventional farming. Liquid manure stored in vats loses vital nutrients, often requiring supplementation with chemical fertilizers; Salatin's method preserves these nutrients, creating healthy compost. Furthermore, the liquid vats produce a noxious smell that permeates the surrounding area; Salatin's approach avoids this offensive odor, benefiting his entire community.

Polyface Farm also uses ponds to capture rain in one season to use for the next. Managing water effectively reduces runoff and erosion and stores enough water to make the land resilient during droughts. The farm doesn't rely on nearby waterways for irrigation, preserving the community's water supply.

While earning a living from his farm, Salatin doesn't do it at the expense of his community, animals, or land. He takes joy in providing healthy food and a clean environment. Unlike corporate farms that feed dead animals and grain to cows, Polyface Farm only provides herbivores with food they would naturally consume. Polyface Farm exemplifies a holistic ecological system where every life plays its intended role, contributing to a complete and healthy cycle.

The farm might sound like a little paradise, but it is nothing new—rather, it is ancient. The rhythms of nature established long before we ever understood them bring about abundance and healing.

Restoring the natural processes of the land took decades. There is no fast track or get-rich-quick pattern in nature. Because of the nurturing process, the Polyface Farm now boasts a higher return on investment than corporate farms, and the nutritional value of its products is significantly greater.

A study by *Mother Earth News* compared eggs produced on Polyface

Farm to those produced on a corporate USDA-approved farm. They found that the nutritional value of the eggs from Polyface Farm significantly exceeded the nutrition of those from monoculture, caged chickens.[6]

	USDA	Polyface Farm
Vitamin E	0.97 mg	7.37 mg
Vitamin A	487 IU	763 IU
Beta-carotene	10 mcg	76.2 mcg
Folate	47 mcg	1,200 mcg
Omega-3s	0.033 g	0.71 g
Cholesterol	423 mg	292 mg
Saturated Fat	3.1 g	2.31 g

Figure 4.1 *Mother Earth News* Study[7]

Rather than focusing on the price, would we not be better off evaluating their nutritional value?[8] Although eggs from chickens pumped full of hormones and crammed into a cage from which they never can forage might be half the price, their nutritional value is less than half. Besides the superior nutritional value, purchasing eggs from natural-raised and fed chickens also supports better treatment of animals. The benefits far outweigh the price difference. To me, opting for the standard eggs from a monoculture farm feels akin to choosing a marshmallow egg over a genuine egg, solely for the sake of a better price. Then again, when we kill off all our chickens and the price of eggs skyrockets, it complicates the choice.

Growing up around friends and relatives who owned farms, I witnessed a rhythm of care and love for animals. Each cow had a name and when one was sick, she would receive special treatment. The cows spent their days in the field and lined up on their own by the gate when it was time to come into the barn for milking. Farms were havens where these domesticated animals lived their best lives.

Then came the day I walked onto my first industrial monoculture farm. In front of me was a cramped pen packed with pigs, their bodies pressed tightly together. The stall owner explained that the pigs were trucked in at a specific weight, confined to this concrete enclosure, and left there until they reached their next target weight. Their feed was designed to fatten them up as they stood, shoulder to shoulder, with no space to wallow.

An overwhelming urge to cry for the pigs and scream at the system washed over me. Disdain boiled up from deep inside me. This was worse than any prison I'd ever seen – at least prisoners have *some* space, *some* connection. How could anyone view this as humane or moral?

I was heartbroken for these intelligent creatures, reduced to mere cogs in a mechanistic system. The system stripped them of their dignity, reducing their existence to efficiency metrics. The entire cycle on these farms feels utterly broken.

It's hard for me to believe that the growth hormones pumped into these animals don't affect the health of Americans trying to maintain their own weight. This approach to farming reflects a mechanistic worldview that sees life as nothing more than machinery.

> As if pigs are no more special than extruded plastic dolls or polyethylene pipe fittings. I would suggest that a culture that views its pigs as just mechanical objects to be reprogrammed and manipulated will view its citizens the same way, and ultimately, God the same way. A deity to be manipulated and formed into something of our liking.[9] ~ Joel Salatin, *The Marvelous Pigness of Pigs.*

We sever cattle from their natural role in land management, opting instead to misuse them for mass production on monoculture farms. This not only results in the land's degradation in their absence, but we also substitute their natural grazing food with grain to expedite their fattening process. We transport grain, fertilized with chemicals, to feed cattle in enclosures filled with manure. A mechanistic view suggests that specialization enhances the bottom line, but it overlooks the extensive destruction left in its wake. Grain feeding is an inefficient land use and an unnatural food for the cattle.

Our responses often complicate the issue. Learning that grain feeding is inefficient, many mistakenly condemn meat consumption, further compounding the problem rather than addressing its root cause. Simplistic arguments like this miss the bigger picture. While grain feeding is unsustainable, grazing is vital for restoring soil health and maintaining ecological balance. Fixing broken patterns requires a nuanced approach that recognizes the interdependencies within ecosystems.

Too often, when attempting to fix a broken pattern, we call everything associated with it evil and break the cycle even further. This was how we ended up culling forty thousand elephants to stop desertification, but instead only exacerbating the issue.

Building something good takes time. The Salatin farm, purchased in 1961, required years of effort to revitalize the land. Now profitable, healthy, and with a potential legacy for future generations, it stands as a testament to proper management practices.

Breaking patterns results in a domino effect of brokenness. Joel Salatins approach to farming gives me hope we can reclaim the rhythmic patterns of life in our food production if we care to.

Chapter Five

The Healing Sanctuary of the Wild

As we moved across the sea, a bit of ocean spray washed over the bow of the small zodiac, misting my jacket. To my right, rugged peaks, heavy with snow and glaciers, rose sharply, their icy masses pressing into the milky waters between us and the shore. Seals and penguins slipped in and out of the surf, leaping effortlessly and breaching the waves. A raft of penguins porpoised beside us, perhaps drawn by curiosity toward the strange visitors in black boats approaching their rookery. They followed, playful and unbothered. Despite their clumsiness on land, they cut through the water with effortless grace. An albatross soared by, and I wondered if any human being could truly gaze upon nature without feeling awe.

I was in South Georgia, a remote island accessible only by sea and only by smaller vessels. It rests just above the South Sandwich islands. The closest inhabited islands are the Falklands, a minimum two-day journey away through the South Atlantic waters. The nearest flight for emergency medical evacuation is also on the Falkland Islands.

South Georgia is a rugged, wild place, hostile to many life forms, but to the birds and sea lions that inhabit it, it is paradise. The island is home to vast colonies of birds, including penguins, albatrosses, and petrels, as well as multiple seal species. The sight of these colonies is something otherworldly.

When we first approached the islands, I stood in awe on the ship's deck, hands resting on the cold metal rail, hoping the weather would permit us to explore these remote shores.

The weather in South Georgia is unpredictable. Trade winds, unimpeded by land masses, hit the island with force, and the high mountains amplify the winds as they funnel through the glacier-filled valleys. Sunshine is rare here, with rain falling approximately three hundred days a year. I felt the growing anticipation among everyone on our expedition.

The day before, we all went through a biosecurity check. We thoroughly inspected every item of clothing, vacuuming jackets, snow and rain pants, gloves, and packs, removing every piece of lint from pockets and seams. We checked all velcro bits for seeds and biomaterials. South Georgia is a protected environment, and any introduced seeds or contaminants pose a tremendous risk to the island. The island is a British territory, and its government is vigilant in guarding against introducing invasive species. Before we left the Falklands a few days earlier, a sniffer dog inspected our ship to ensure there were no rats on board.

Cruise ships carrying more than 850 passengers are not permitted in the waters near South Georgia. Most landing sites are open only to vessels with fewer than 200 passengers. We were on a smaller vessel and hoped to visit as many of these landing sites as possible. Our goal was to observe and learn, while leaving no trace behind. Minimizing the disruption of our presence was a priority. At night, the crew extinguished all outside deck lights and instructed passengers to close their blackout drapes over the portholes in their cabins. This would minimize light pollution and avoid confusing the birds.

We dropped anchor in Right Whale Bay. After putting on layers of cold-weather gear and rain protection, we prepared to board the zodiacs. Each of us carefully stepped into a tray of disinfectant to ensure our Wellington boots were free of contaminants—a process we repeated before and after every landing. After disinfecting, we made our way down to the gangway and climbed into the zodiacs.

The sky was gray and fog hung low over the bay, adding an air of mystery to the shore where we landed. Penguins and seals watched as we disembarked in the surf and waded onto the black beach. The fog seemed to intensify the colors of the summer tussock grass growing in clumps on the hills. Green vegetation covered the base of the mountains, while snow adorned the peaks and formed flowing veins of white down the crevices. A waterfall descended from a cliff to the north. It felt like stepping back in time, and I half-expected a prehistoric creature to emerge from the mist.

Seals and penguins populated the beach. We walked among them as careful observers, keeping a respectful distance. The wildlife paid little attention to us, except for the occasional territorial fur seal that made its claim to the beach known. Here, wildlife have the right of way, and we tried to stay out of their paths. Baby seals gathered in a small pool at the base of a large rock outcropping, playing while their parents hunted for food at sea. They frolicked and splashed and occasionally chased a human observer.

The penguin colony extended up the hillside, each tuxedoed bird blending into the masses, creating a pattern of black, white, and brown atop the green slope. I had always dreamed of seeing far-off places, and here I was, standing on a remote island far from civilization, engulfed in mist and mystery, a world away from anything I had ever known. It humbled me.

After exploring a few landings on the west side of the island, we sailed around to Stromness Bay. A few buildings remain there, remnants of a whaling station built in 1912 and in operation until 1931.

It was used as a ship repair yard for a while but was abandoned by 1961. The remaining structures stand in a state of disrepair, the red earthen rust blending into the raw mountainside behind them, like ghosts of the past.

We viewed the buildings from a distance—remaining asbestos kept us from getting too close—and discussed early explorers like Ernest Shackleton, who found rescue here after his ship was trapped in Antarctic ice. Most of his crew waited on Elephant Island near Antarctica, while Shackleton and five others made an 800-mile (1,300 km) journey in a lifeboat across treacherous seas to South Georgia in search of rescue. They landed on the east side of the island, then navigated on foot over the glacier-capped mountains to the whaling station. Those were the days when exploration meant venturing into the unknown.

We anchored in Stromness Bay on a calm, sunny day with only a few clouds—a rare occasion in South Georgia that we fully soaked in. The whaling station was on the edge of a verdant green valley surrounded by snow-capped mountains. I wandered to the farthest end of the valley and gazed at the towering peaks, longing to follow the trails marked on my map, but the limited time on these protected shores prevented me from exploring further.

Our next landing was in St. Andrews Bay where another beautiful day greeted us and the weather seemed to be in our favor. We swung our legs over the zodiacs and jumped into the surf to wade toward shore.

On land, we navigated around elephant seals, fur seals, and penguins scattered along the beach, heading toward a penguin colony that filled the valley to the north. The call of King Penguins raising their trumpets to the air rose above the sound of the wind. Thousands of penguins stood before me, calling to their mates and young. The young penguins waddled around us while their brown, downy feathers rippled softly in the wind. One sidled up to a friend next to me and began pecking at him, while another chick with its side-to-side gait chased a Skua bird away, as curious and playful as any child.

In the afternoon, we climbed a small hill above the penguin colony, following a grassy slope beneath towering glaciers that cascaded from a valley of rugged, snow-capped mountains. On the hill, I looked down, and the scenery overwhelmed my senses. My eyes could scarcely take in the sea of penguins spread across the valley below. Over 150,000 breeding pairs nest there, and their trumpeting calls echoed in my ears. My sense of smell took in a mix of penguin stench and the fresh breeze off the glaciers behind us. We all stood there, smiling and shaking our heads in wonder at the vast numbers filling the valley and spilling out onto the beach.

Farther down the beach, fur seals and elephant seals wove through the more dispersed penguins. Seals are protective of their space. If we crossed their invisible boundary, they made it known with a deep, belch-like growl. They would charge if we didn't heed their warning, so we always moved back from their boundary. Fur seals charge faster than most other seals, but still slowly enough for us to respond. We were successful at avoiding confrontations, but we knew that if a seal was surprised, we should respond as we would to a bear: stand our ground, face it, and raise our arms to appear larger (though we were never larger). We always gave them the right of way.

Grytviken, just north of St. Andrews Bay, is the only place in South Georgia with a working settlement. While it no longer has permanent residents, there are usually eleven to thirty people there throughout the year, including researchers and temporary summer staff to manage the South Georgia Museum. This is the only bay around South Georgia where passengers from larger ships may disembark. A major attraction is the Grytviken graveyard, the final resting place of polar explorers like Ernest Shackleton and Frank Wild.

We were fortunate that the weather on our trip allowed for multiple landings each day. We knew we were incredibly lucky to have seen so much. Each landing in South Georgia deepened my sense of the island's remoteness and wildness. This untamed solitude stirred something deep within me—a quiet exhilaration, a longing to be fully enveloped in its beauty.

However, we missed our landing in Grytviken because of high winds so extreme they kicked up a waterspout. While waiting for the winds to die down, our captain took the ship into Moraine Fjord, giving us a closer look at the glaciers pushing into the sea. But the winds continued to blow unabated, making a landing in Grytviken impossible.

As we sailed away from South Georgia, I made my way to the open top deck. This was my favorite place on this ship—often quiet, with few to no other people around. I gazed out at the silhouette of the mountains standing alone in the vast ocean. The sun was low in the sky, and color filled the horizon. A strong wind whistled through the bolt holes of the glass screen across the front of the deck. A wandering albatross with wings spread wide glided effortlessly on the winds. Then it rose gracefully and swooped down over the water. I felt the peace of the unseen wind sustaining its path.

The sea seemed to call my name, whispering of places unknown, of depths beyond imagination yet to be explored. This unknown compels me to wander further, yet I am not lost. This is where I feel most at home—amid the unknown, in the open spaces of nature that proclaim majesty. During these moments, the important things come into focus, and the rest lose their fight for space in my mind.

Scientists have not yet discovered the exact reasons time in nature seems to bring peace and healing. Perhaps it is the clean air, the changes in noise levels, the smells of the forest, the energy of the plants and trees, or the natural colors of the blue ocean or the green vegetation. Maybe it is the pattern of it all that calms us. Regardless of the reason, nature's beauty can fill us with joy. Scientists study its patterns to uncover the truths it holds.

Most people claim that time in nature makes them feel better. While this sentiment is subjective, the fact that a large enough percentage of people feel this way points to a truth we just aren't able to nail down into a formula yet; it is still beyond mathematics. There is something in the peacefulness of the woods or the immensity of the sea that forms an intangible yet often felt rhythm.

Science Direct research journal asserts that numerous studies increasingly show that spending time in nature can produce tangible benefits for both mental and physical health, along with a variety of other positive outcomes.[1]

Those of us who have spent time in the forest can understand and appreciate *Shinrin-yoku*. The Japanese Ministry of Agriculture, Forestry, and Fisheries created the term *Shinrin-yoku,* meaning "forest bathing." The ministry recommends "absorbing the forest atmosphere" by resting in the forest, smelling the fragrance of the flora and fauna, breathing in the clean air, and gazing at the lush and serene abundance of life.

Forest bathing might sound strange to the Western mind initially, but sunbathing sounds absolutely normal. Each year, thousands of people flock to white sandy beaches to soak in the sun's heat and breathe in the fresh salty air as they "sunbathe." The pattern of ocean waves lulls us to slow down and listen, providing a peaceful rhythm along the shore. Even thinking about it can be relaxing.

Forest bathing is a similar concept, but it happens in the woods instead of on the beach. Instead of salty air, it might be the fragrance of pine we breathe in. Instead of baking in the sun, we might cool off in the air kept temperate by the trees. On the beach, the rhythmic sound of the waves is a lullaby. In the forest, the sound of leaves dancing in the wind calms our heart. Rhythms attract us; they soothe us, even if we can't scientifically measure why.

A study by the University of East Anglia summarized their research of the health boost from nature saying:

> Living close to nature and spending time outside has significant and wide-ranging health benefits—according to new research. A new report reveals that exposure to greenspace reduces the risk of type II diabetes, cardiovascular disease, premature death, preterm birth, stress, and high blood pressure.[2]

Peter Wohlleben, in his book *The Heartbeat of Trees*, highlights that longing for nature is an instinct.

> Is it any wonder that nature constantly calls us out of the cities and back to our roots? Isn't this a healthy instinct, which shows that we are still completely in touch with our senses? The people who call this escapism are the ones who are out of touch.[3] ~ Peter Wohlleben

Science has yet to prove conclusively that time in nature is healing, but the growing evidence points us toward a deeper truth. Those who have felt the joy of nature's majesty will instinctively recognize that something profound is at work. There is a connection, a rhythm—an invitation to understand more.

No organism can survive in complete isolation. I cannot heal myself. Yet, as I stood on the ship deck leaving South Georgia, watching the sun sink behind snow-capped peaks rising from the ocean, it felt like a remedy for my soul. This beauty is a gift freely given—one we need only to receive.

The Symmetry of Nature

Beauty and symmetry effortlessly traverse the boundaries among all scientific disciplines, as if they are melodies that bind these diverse rhythms together.

Patterns of symmetry lead physicists to truth.
They use symmetry as a guiding principle to uncover the laws of the universe.

Even in chaos, patterns emerge, revealing rhythms in seemingly unpredictable outcomes. Scientists have discovered that what may appear random is often governed by slight variations that lead to unpredictability and shape the future in ways we could not foresee.

CHAPTER SIX

Nature's Rhythms and Design

THE SUN, MOON, STARS, and gravitational forces bestow upon us the cycles of daily life and the changing seasons. They grace us with splendor in the vast night sky. Each morning, the sun orchestrates a magnificent transformation, painting the sky with its radiance and casting warmth upon the landscape. As we study distant celestial objects in our universe, we strive to unravel the laws that govern them, delving into the realms of quantum mechanics. Experimenting with quasars light years away, we test the principles of quantum entanglement.[1] It is almost as if the answers are inscribed in the stars. The patterns and laws we uncover become even more captivating when we recognize the same harmonious patterns beside us and all around us on earth.

> Einstein's great advance in 1905 was to put symmetry first, to regard the symmetry principle as the primary feature of nature that constrains the allowable dynamical laws. . . . With the development of quantum mechanics in the 1920s, symmetry principles came to play an even more fundamental role. In the latter half of the 20th century, symmetry has been the most dominant concept in the

exploration and formulation of the fundamental laws of physics. Today it serves as a guiding principle in the search for further unification and progress.[2] ~ D.J. Gross, *Proceedings of the National Academy of Sciences*

Every major advance in physics for more than a century has turned on revelations about symmetry. It's there at the dawn of general relativity, in the birth of the Standard Model, in the hunt for the Higgs.[3] ~ Kevin Hartnett, *Quanta Magazine*

Science serves as the tool to articulate our observations in the natural world, and mathematics stands as its language. But symmetry guides scientific discovery, and the experience of beauty is beyond its limit.

Fading Footprints in the Oasis

Trin and I awoke in a small hut in Atins bathed in diffused sunlight. The thin walls, made of white corrugated plastic only slightly stronger than cardboard, did little to keep out the brilliance of the rising Brazilian sun. They seemed fragile enough to crumble if I stumbled against them. Whisky, a local dog, rested on the small porch just outside our door, as if standing guard through the night.

From our 15x15-foot room, we descended rickety stairs crafted from driftwood. Below our hut, which was raised on stilts, a small kitchenette was open on three sides to the warm air of Lençóis Maranhenses National Park. We had arrived just yesterday, traveling by jeep along sandy roads pockmarked with water-filled depressions—some shallow, others deep. The town is accessible only by boat or four-wheel-drive vehicle.

We made coffee and packed a bag with lunch, water, snacks, and sunblock. Our sole agenda for the day was to explore the dunes just outside our doorstep. Within a mile of our hut, our feet sank into brilliant white dunes that stretched endlessly under the vast blue sky. Strong winds carried granules of sand, creating a misty waterfall effect over each ridge, blurring the downward slopes.

Frequently, a valley surprised us with a pool of clear water resting in its basin. The clay beneath the sand in this region retains water from the rainy season well into summer, forming pristine lagoons that mirrored the sky above. We descended into one such valley, sliding on sand that felt like finely sifted flour. The water cooled and refreshed us as we immersed ourselves in its tranquil depths. After one rejuvenating dip, we sat beside a lagoon, savoring our packed lunch while listening to the wind gently reshaping the landscape before us.

For days, we wandered through the park, drawn to the water's edge where soft, silty ground tugged at our ankles. Sometimes, we took a few extra steps to experience the playful grip of quicksand, daring us to explore further before we pulled ourselves free. Other times, we stood captivated by the gentle undulations of the pristine white hills—a sandy landscape in perpetual motion, exuding perfect tranquility.

"I think this is the most beautiful place I have seen," Trin said to me.

I nodded in agreement, my eyes brimming with tears of joy at this extraordinary display. Footprints trailed behind us as we walked toward the horizon. Within moments, the wind erased all traces of our journey, reclaiming its patterns in the sand. The memory remains vivid in my soul.

Wind and dancing grains of sand create patterns that grace dunes and beaches worldwide, shaping the landscapes and even the ocean floor. Across the continent, on Chile's Pacific coast, the expansive dunes of Moon Valley of the Atacama Desert showcase this distinctive wind ripple pattern in their mocha velvet sand, shimmering under the sun. Warm grains of the Atacama felt like satin as they sifted through our fingers. On the far side of the globe in the Sahara's Erg Chebbi, sand dunes glow with a reddish hue as the sun descends.

Many places have held me transfixed with their magnificence, even in harsh environments, like Lençóis Maranhenses, the Sahara, and the Atacama Desert, where life struggles to exist. Wind-sculpted valleys and muted hues of sand create captivating artwork. Each of these sandy deserts is unique—like snowflakes and fingerprints—yet as we sat atop their dunes, the patterns in the sand seemed to echo one another, as though forming a symmetry across the globe.

The exploration of these repetitions in nature is limitless, and their splendor is astounding. Recognizing these designs offers insight into our world. Symmetry, a distinct form of pattern, serves as a key that unlocks many principles in physics, revealing order beneath the surface of our ever-changing world.

Another pattern, both mathematically defined and echoed across our planet, is worth noting. Let's dive into the numbers—just for a moment. If numbers aren't your thing, stay with me; this will be brief, but the patterns they reveal are truly fascinating.

Spirals of the Sunflower

> "Little flower—but if I could understand
> What you are, root and all, and all in all,
> I should know what God and man is."[4]
> ~ Alfred Tennyson

Sunflower seeds, positioned on the face of the flower and turned toward the sun, arrange themselves with optimal space efficiency, following a pattern known as the Fibonacci sequence.

The numerical Fibonacci sequence is formed by adding two numbers to get the subsequent number. Adding the first two numbers below (0+1) yields one. The next two (1+1) results in two.

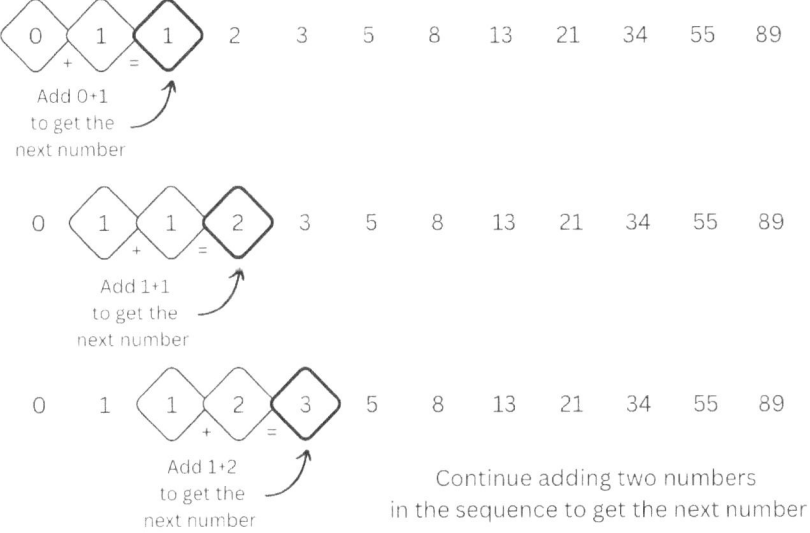

Figure 6.1 Fibonacci Sequence

The elegance of this progression becomes more apparent when observing the proportions between numbers, as shown in the graph above. To visualize this, we construct a square corresponding to the first number in the sequence, then continue adding boxes in a circular arrangement, each representing the next value. Once all the boxes are in place, connecting the opposite corners of each square with an arc reveals a spiral—familiarly seen in the cross-section of a nautilus shell.

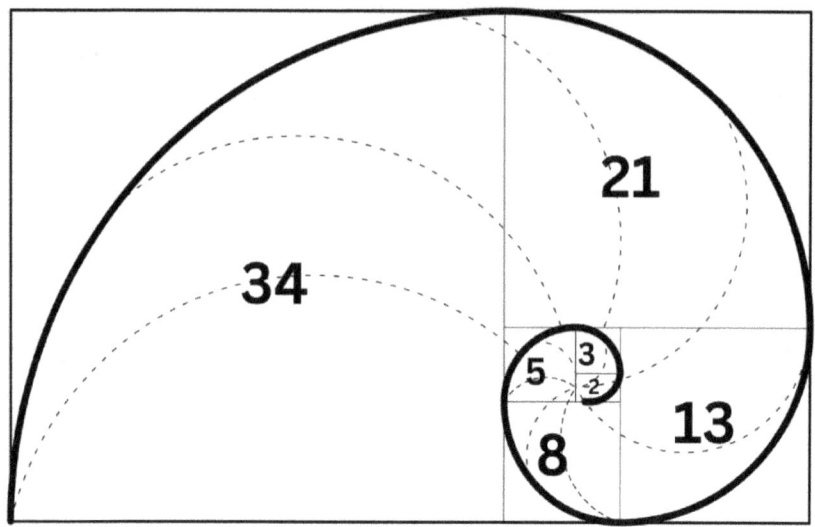

Figure 6.2 Fibonacci Pattern[5]

This logarithmic spiral is called the Fibonacci pattern. Not only do we see this distinctive design in the seeds of a sunflower and the cross-section of Nautilus shells, but it also appears in pine cones, ferns, roses, cauliflower, and the segmental arrangement of pineapples. Remarkably, it also appears in the formation of hurricanes and distant galaxies. Symmetry and repeating patterns are prevalent in nature.

Within the Fibonacci sequence, the ratio of each number to the preceding one tends toward ɸ (phi), pronounced "fee" or "fi," (like the "Fee-fi-fo-fum" from *Jack and the Beanstalk*). The digits to the right of the decimal point extend infinitely, never repeating, rendering ɸ an irrational number. Commonly referred to as the golden ratio, ɸ has earned nicknames such as the "golden number" or "nature's code."

You have undoubtedly heard of π (pi, 3.14159 . . .), another famous irrational number. π, pronounced "pie" (like something good to eat), is the ratio of the circle's circumference to its diameter. In comparison, ɸ (1.61803 . . .) is slightly more complex, but we can calculate it geometrically by showing relationships within various shapes, including lines and pentagons.

When nature forms seeds, petals, or leaves based on the ratio of ϕ, it maximizes sunlight exposure and minimizes overlap. This efficient design allows seeds, like those in a sunflower, to be packed into a compact space. Similarly, the arrangement of leaves and petals optimizes water collection and sunlight absorption, enhancing the plant's ability to thrive.

Plants, of course, do not possess mathematical intelligence. Internal hormones direct new growth, generating these patterns as they develop. This process is inherent in their genetic makeup, serving both functional and aesthetic purposes. It's important to emphasize that their efficiency stems from inherited genetic information—it's not something plants figure out; their genetic blueprint directs this development.

In the past, people revered ϕ as a mysterious and wonderful proportion, giving it titles such as the "divine proportion" because of its association with beauty in nature, art, and architecture.[6] Craftsmen used this ratio to make musical instruments like the Stradivari violin. Architectural marvels such as the Egyptian Pyramids, the Pantheon, Notre Dame, and the Taj Mahal, as well as modern creations like Lego's experience center, incorporate the ϕ ratio in their designs.

Human creativity often mimics the natural world, drawing inspiration from its patterns and rhythms to shape our own unique expressions.

The golden ratio (ϕ) also plays a role in creating optical illusions that induce a false sense of movement. Artists and architects incorporate this mathematical proportion into their designs, creating an allure that captivates the eye. Leonardo da Vinci's sketches detail his use of calculations based on ϕ, while Salvador Dalí used the golden ratio (ϕ) in his masterpiece *The Sacrament of the Last Supper*.

While we may not consider the ϕ ratio or the Fibonacci sequence to be mystical today, it seems to serve as a connection between our logical, mathematical brains and our emotion in the perception of beauty.

It's easy to fall into the trap of viewing logic and emotion as entirely separate, disconnected entities. I occasionally reflect on the stereotype of a mathematician, picturing someone like Spock from Star Trek—purely logical, devoid of emotion—perhaps because he was my childhood hero. Conversely, artists are often perceived as deeply emotional. It promotes the idea that mathematicians and artists belong in separate boxes, like the separation we often place on science and faith (more on that separation later). In reality, logic and emotion are not opposing forces; they constantly influence each other within every individual. The natural world embodies this balance, seamlessly blending beauty and the emotion it evokes with beautiful mathematical precision. It reminds us that these elements are not only connected—they thrive together.

The Language of Numbers

We convert aesthetically pleasing patterns into numbers because math is the language used to describe the order we see in the universe. Basic math has been part of communication among humans for as long as we can tell. Complex math developed beside our quest to recreate patterns from nature and understand more of our universe. The Egyptians used trigonometry to build the pyramids and the Greeks used it to trace the movement of the stars.

Our analytical minds find comfort in numbers, thinking that if we accompany an idea with a calculation or formula, we can know it with certainty.

Humans like order. Other creatures might enjoy it too, though we can't speak to their desires. Patterns attract us; we find their elegance appealing. We are inquisitive creatures who seek explanations for those patterns, desiring to understand how they developed and interact with each other. This curiosity draws us in, tantalizing our desire for understanding.

Quantum Field Theory is on a quest to quantify and comprehend our universe using the common language of mathematics. Mathematics holds the potential to answer many of the still-unanswered questions within this discipline, but it cannot answer all of them; science can describe, but it cannot answer why. Scientific discoveries are fascinating, sometimes mind-blowing, and always exciting, but there is always the unknown that we must contend with.

CHAPTER SEVEN

Symmetry—The Physicist's Key to the Universe

PHYSICISTS TODAY ASSERT THAT "symmetry lies at the heart of the universe and appears to be the fundamental property."[1] Arvin Ash, a chemical and mathematical engineer, stated, "If you ask a physicist what makes the universe tick or what is at the core of physics theories, the most common answer would be symmetry."[2]

The two leading labs dedicated to particle acceleration in the USA call their joint magazine *Symmetry*. The byline for their website is: "Symmetry is your view into the world of particle physics."[3]

Emmy Noether was a German mathematician whom Albert Einstein referred to as "the most significant creative mathematical genius thus far produced since the higher education of women began."[4] She developed Noether's First Theorem, which reveals the value of physical symmetries and their importance in discovering the laws of conservation. Noether's Theorem states that for every constant symmetry in physics, there is a corresponding conservation law.

Think of symmetry as something that stays constant, like a well-balanced seesaw. Noether's Theorem tells us that every time you notice such a balance in nature, there's a hidden conservation law keeping everything in check. For example, the unchanging flow of time ensures that energy isn't magically Created or destroyed—it's conserved.

Symmetry has been central to discovering the universe's fundamental forces, like gravity and electromagnetism, helping scientists uncover some of the most basic rules that the universe seems to obey.

It is more than just a tool for science; symmetry resonates deeply with us. We find it beautiful and reassuring because it introduces a sense of order, making the world feel more predictable. This innate appreciation may be more than just aesthetic—it could be an instinct that drives us to explore, question, and seek understanding in the patterns of the universe.

Symmetry inspires the things that humans create. A few examples of innovations that copy nature include Velcro, inspired by burdock plants;[5] Geckskin, derived from the gecko's grip;[6] and riblets on aircraft, increasing aerodynamic lift by 323%, copied from a shark's denticles.[7] We don't truly Create[8]—make something from nothing—we merely copy from the natural world.

Wanderers in an Earth-Centered Universe

Before the sixteenth century, scientists believed in the geocentric model, asserting that Earth was the center of the universe. The sun and planets were thought to revolve around the Earth. The Greeks termed other planets "wanderers" because of their seemingly chaotic paths through the sky.

In 1515, Nicolaus Copernicus proposed a revolutionary heliocentric model, positioning the sun at the center of the universe. It took nearly a century for this idea to gain acceptance. If we had used today's standards of determining truth, involving peer reviews and consensus, we would have deemed Copernicus "wrong" and considered his ideas dangerous. Spoiler alert: this is precisely what happened. Consensus "proved" Copernicus wrong and labeled his heliocentric model as dangerous. Things have changed little over time.

The challenge of letting go of the entrenched belief that Earth was at the universe's center was not merely because of the difficulty of change itself; it also involved relinquishing the sense of grandeur associated with being the center of everything. Giordano Bruno was burned at the stake for teaching Copernicus's heliocentric view, among other "heresies." He was not alone in facing persecution for challenging the accepted beliefs of both the scientific establishment and the Church.

The church was the ruling power of the time that sought to guard truth, similar to politics today guarding against "misinformation."

Evidence supporting the heliocentric model grew when Galileo observed the phases of Venus, providing further proof that the Earth revolves around the sun. Although Copernicus had published his heliocentric model in 1543, most of the educated world still clung to the geocentric view. The Catholic Church, the acting political authority of the time, even declared in her final judgment, "This proposition receives the same judgment in philosophy."[9] During the seventeenth century, philosophers played the role of scientists.

In the early seventeenth century, Johannes Kepler's calculations on planetary orbital movements, based on the Copernican model, led to the formulation of the laws of planetary motion. Kepler's elliptical calculations dispelled the notion planets wander. What was once thought to be chaotic and random now revealed a clear pattern—further confirming the heliocentric model. It wasn't until later in the seventeenth century Europe finally accepted this revolutionary view.[10]

Only after we let go of the belief we reside at the center of the universe could we comprehend the movements of other planets. Discovering Truth enhances our understanding and diminishes what once appeared chaotic. By realizing Earth's position in the universe, we gained the ability to mathematically calculate the predictable pattern planets follow.

Later, Einstein's theories of relativity revealed our personal perspective is not the center of reality, further refining our understanding of our place in the universe. When we understand our place, things around us make more sense.

Many people might perceive the world as the product of randomness, yet chaos theory introduces a fascinating perspective. Rather than a disorderly jumble, this theory suggests unpredictability can possess an underlying pattern.

CHAPTER EIGHT

The Pattern of Uncertainty

TRAVELING ON CHICKEN BUSES in Nicaragua can be an unpredictable adventure. These repurposed American school buses serve as public transportation throughout Central America, but Nicaraguans transform them into something unique. Vibrant colors, religious symbols, and metal luggage racks—both inside and on the roof—give them a distinctive look. Passengers' belongings range from produce to mattresses, all secured tightly on top.

The bus only reaches maximum capacity when no more bodies can be squeezed into the aisle or stairwell. It's not uncommon to see arms, heads, and sometimes even feet dangling out of windows and doors because of the lack of space inside. Sometimes, Trin and I rode next to live poultry. While the origin of the term "chicken bus" is uncertain, actual chickens and other livestock onboard may have contributed to its name. These animals are often, amusingly, the quietest passengers, contrasting with the lively conversations that can extend well into the early hours of the morning on an overnight journey.

We enjoyed riding local transportation because it immersed us in the daily rhythms of Nicaraguan life, offering a firsthand glimpse into the culture. Each journey was a unique experience—sometimes pleasant, sometimes chaotic, and occasionally even dangerous.

More than once, we received a text from friends checking in on us from the USA, who heard about a bus careening off the edge of a cliff in a country where we were traveling. Many times, these types of accidents result in large death tolls. One devastating crash killed all fifty-nine passengers aboard. Just two weeks earlier, we had taken that same route, rounding that same treacherous bend.

Yet, beyond the risks of mountain roads, simply getting on and off could be its own adventure. One particularly memorable trip was from Jinotega to Matagalpa. When the bus pulled up, there was a frenzy as everyone scrambled to board, eager to squeeze themselves in. By the time we boarded, every seat was full and bodies were pressing close in the already packed aisle. Then I felt it—a hand exploring my pockets and a tug on the zipper of my pack. I turned to see a man wearing a Cheerios shirt with the logo slightly off-center; the shirt had probably been sent here from the USA when it didn't pass inspection. I told him no and shoved him to maneuver away from his prying hands.

Meanwhile, farther down the aisle, Trin was also experiencing someone probing his pockets. Thankfully, nothing of value is in our pockets, and we secured all the important zippers on our backpack with padlocks. The one item in my pocket was a napkin I had just blown my nose into; unfortunately, the man didn't take that.

The attendant insisted we hand over our packs to be stowed on the roof rack. We refused, knowing the chances of our packs returning intact and with everything inside were slim. Normally, a simple no suffices, and the matter is dropped, but this guy was persistent. Removing our packs would free up space for two more passengers. Understanding this, we got off the bus to wait for the next one.

Thirty minutes later, the next bus to Matagalpa arrived. Like before, the crowd swarmed the curb. Luckily, this time, the rear door stopped near me. I joined the throng and secured a seat in the back. Occupying the entire seat, I awaited Trin's arrival. Once he joined me, we settled in with our packs resting on our laps, ensuring they didn't occupy space another passenger could use. Despite the already crowded conditions, more passengers squeezed in, occupying every available crevice.

As the vehicle lurched forward, Trin and I exchanged amused glances. Despite the momentary frenzy, we had successfully navigated the journey "Nica style," embracing the unique rhythm of Nicaraguan transportation. What seemed disorderly at first revealed an unspoken system—one that, in its own way, ensured everyone eventually reached their respective destination.

Our universe—and our lives—often feel chaotic. We may not know what will happen next or why certain events unfold the way they do. Yet, chaos theory suggests even in unpredictability, patterns emerge. What seems like randomness may have an underlying order.

Take the packed chaos of a chicken bus, for example—no matter how overwhelming the crush of people, we always reached our destination, often with a great story to tell. Was it mere chance, or is there something more finely tuned about our world? Could there be an unseen force shaping what we perceive as disorder?

The Butterfly Effect

In the 1960s, meteorologist Edward Lorenz, a mathematician at heart, argued weather was fundamentally unpredictable due to its sensitivity to initial conditions. He described how slight variations—such as the way sunlight warms the angle of a rock—could alter heat convection, influencing wind patterns and, ultimately, the weather itself. With countless tiny factors shaping atmospheric change, precise long-term forecasting remains elusive.

Lorenz coined the term "butterfly effect" to illustrate this sensitive dependence on initial conditions. Miniscule changes within his weather experiments changed the outcome's trajectory.

In 2004, the sci-fi movie *The Butterfly Effect* was released, and loosely used this principle to suggest small changes in the past could drastically alter one's life. It's an entertaining and whimsical illustration of how Lorenz's findings on initial conditions can have a large impact.

Envisioning a massive effect from the delicate flap of a butterfly wing seems implausible, but if we look at this mathematically using a financial investment example, the theory Lorenz discovered makes more concrete sense.

A $50 initial one-time investment, compounded monthly at a 10% annual interest rate over 50 years, will grow to $7,269. Interestingly, if we slightly alter the initial condition by increasing $50 to $51, the accumulated results are $145 higher after 50 years. Expanding this scenario to 100 years widens the gap even more, with the earnings from the initial $1 difference totaling $21,132. This underscores the significant impact small initial conditions can have on the result, particularly as the trajectory extends over a longer period.[1]

Initial Investment	Value after 10 Years	Value after 100 years
$50	$7,269	$1,056,621
$51	$7,414	$1,077,753
One dollar difference in initial condition	One dollar difference in initial condition added up to a **$145 difference** after 10 years.	One dollar difference in initial condition added up to a **$21,132 difference** after 100 years.

Figure 8.1 Initial Investment

In recent decades, we have made significant strides in advancing our ability to issue timely warnings for flash floods and major storms. Using supercomputers aggregating data from numerous sensors measuring wind, rain, barometric pressure, and heat, we gain valuable insights into weather patterns and can track storm systems evolving over the ocean. This extensive data collection enhances our predictive capabilities. Despite these advancements, myriad factors influencing the weather's inherent complexity limit our predictive precision. Accuracy diminishes notably after a mere ten days.

Improved understanding of cause and effect, fueled by continuous discoveries, has enhanced weather forecasting. Countless data points, however, still surpass our tracking capabilities. Observing seasonal patterns provides valuable clues. For instance, mid-summer in Phoenix, Arizona, is reliably sunny and scorching hot, while December in upstate New York typically demands a warm coat. Yet, predicting cloud cover one day in advance in the Scottish Highlands remains a challenge.

Our world operates within a complex, interdependent system. As we explore its intricacies, it becomes clear what may initially seem chaotic often reveals underlying order and patterns. While there is still much we can't predict—due to our inability to track all the initial conditions—these patterns suggest order exists. The full scope of cause and effect, however, lies beyond our ability to track and exceeds the small fragment of knowledge we hold in the vastness of the universe.

Strange Attractors and Fractals

Around the same time Lorenz was proclaiming weather's unpredictability due to initial conditions, mathematician Stephen Smale was researching dynamical systems. Smale attributed his experiments' unpredictable results to "strange attractors," creating a dynamic he believed defied prediction. He characterized these dynamics as "chaotic."

Researchers later discovered these "strange attractors" possess detailed structures when examined closely under magnification.

The Barnsley fern serves as an excellent illustration of detailed structures. The fern blade, depicted below, assumes a triangular shape. Each pinna, extending symmetrically on both sides from the central stalk, mirrors the blade's shape but on a smaller scale. Upon closer examination, each segment, a tiny section growing symmetrically from each pinna, retains the same shape but on an even smaller scale. The fern consistently reproduces this pattern on progressively smaller scales. Scientists refer to this repeating pattern as a fractal.

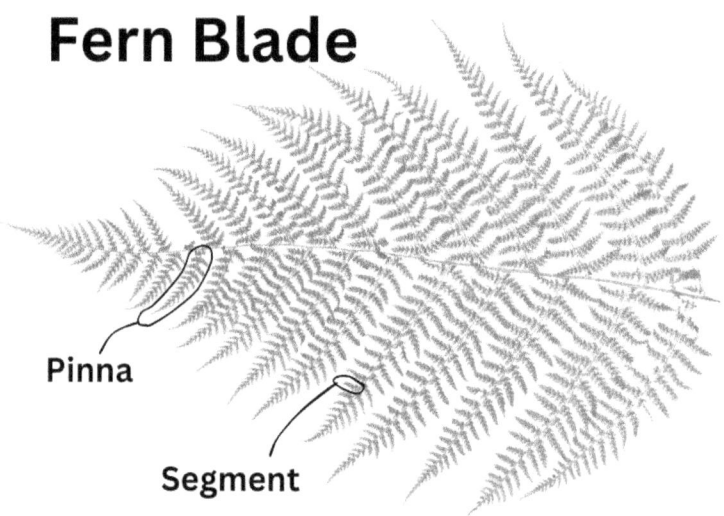

Fern Blade

Pinna

Segment

Figure 8.2 Barnsley Fern[2]

The unpredictability in Smale's results stemmed from an imperceptible detail. Unraveling the mystery behind these "unpredictable" outcomes unveiled the realm of fractals, alternatively recognized as unfolding or expanding symmetry.

Fractals, as exemplified by the Barnsley fern, are characterized by self-similarity (a specific type of symmetry), meaning the same patterns repeat at different scales.

Nature showcases many instances of fractal patterns, spanning across plants, trees, waterways, seashells, and even mountains. The skies above us feature fractals in the guise of clouds and hurricanes.

We even emulate this natural phenomenon in our music. A musical fractal is a composition where the theme harmonizes with a slowed-down version of itself. Johann Sebastian Bach masterfully explored symmetry and fractals in his fugues.

Unveiling the world of fractals has shown us what we once perceived as chaos actually hides a profound order. It has exposed more patterns of the universe that repeat themselves in progressively smaller scales. It shows us the reason our predictions are inaccurate, the cause being unknown initial conditions.

An example of fractals influencing our predictions occurs on the putting green. A skilled golfer might calculate the perfect angle and force needed for a hole-in-one. Yet, a tiny stone on the green could cause the ball to veer off course and miss the hole. What Smale discovered were angles and shapes that were invisible or unmeasured. These fractals are conditions that alter the course of predictions. This insight transformed determinism, the belief everything is preordained, from a purely theological debate into a topic of scientific inquiry.

In 1814, French scholar Pierre-Simon Laplace presented a paper titled "Laplace's Demon," delving into the realm of determinism. He proposed that if one possessed exhaustive knowledge of the universe, including the minutiae of every atom, the angle of every fractal, and every initial condition, one could predict anything—knowing everything. Laplace wrote,

> It [an intelligence sufficiently vast] would embrace in the same formula the movements of the greatest bodies of the universe and those of the lightest atom; for it, nothing

would be uncertain and the future, as the past, would be present to its eyes.[3]

Many have disputed Laplace's ideas, but the debate over determinism, with its theological, psychological, and scientific implications, traces back to some of the oldest-known written texts.[4] Free will vs. determinism is not something our current scientific method can formulate to mathematical precision.

Determinism suggests every atom in our universe follows a predetermined course. Often linked to concepts like fate or predestination, it fuels debate over the tension between choice and inevitability. On the other end of the spectrum is the idea that chaos reigns—that the future is unknowable, even to Laplace's demon. In practical terms, if nothing is predetermined, then our choices—our free will—can shape our lives entirely; we are self-made, with no overarching plan guiding us.

Perhaps, however, determinism and free will coexist. Agency may operate within a framework of predefined parameters and constraints—a framework that stretches beyond our scientific measurement. This parallels the intricate yet unpredictable patterns of Lorenz's waterwheels, which we'll explore shortly.

Ultimately, the true nature of this debate lies beyond the reach of current scientific inquiry. Pure determinism and absolute autonomy exist as extremes, but in reality, most of us navigate life embracing elements of both, whether consciously or not.

Still, figures like Michio Kaku, co-founder of string theory, place their trust in mathematical knowledge as the ultimate solution. In his book *Quantum Supremacy: How the Quantum Computer Revolution Will Change Everything*, Kaku articulates his faith that quantum computers hold the key to resolving all of the world's problems. He suggests with quantum computing, we could track all conditions and thus improve on nature, fix all failures, and predict the future.[5]

Despite Kaku's faith, knowing the initial condition of every atom in the universe since the beginning of time is an impossibility for humankind. Even with advancements in quantum computing, the required data sets from the past to predict the future are unattainable. While we can make educated guesses about the initial conditions of the past, obtaining precise historical figures—something seemingly simple, such as the exact number of people on Earth each year—is unattainable and lost in history, even for those years with written records. If we possessed instruments capable of measuring every known atom that could feed into a quantum computer, we would still lack information about initial historical conditions. Scientists estimate a quantum computer capable of measuring every data point would need to be as vast as the universe itself.

The actualization of humans attaining the status of Laplace's demon, irrespective of the number of qubits available for computation, surpasses the limits of achievable knowledge.

When Random Becomes Remarkable

Lorenz, hailed as the pioneer of chaos theory, unearthed something even more intriguing than Smale fractals within his groundbreaking work. Despite proving through multiple experiments that prediction is an elusive task, Lorenz's findings unveiled underlying patterns.

Figure 8.3 Malkus Waterwheel[6]

In an experiment involving waterwheels, also known as the Malkus waterwheel, two symmetrical wheels held buckets equipped with small holes for water release. The buckets received a flow of water from above with a volume greater than the small holes' release capacity.

As a result, the buckets began filling, setting the wheels into motion. Periodically, the wheels came to a halt, only to resume when the uppermost bucket reached full capacity. Subsequently, the wheel exhibited erratic turns, alternating between clockwise and counterclockwise directions. The wheel's speed varied unpredictably, and both the movements and velocity defied mathematical prediction. The overall result presented seemingly random behavior, with identical wheels turning in unpredictable directions.

Lorenz didn't conclude his exploration of the waterwheel there. He extended his analysis by plotting the movements in three dimensions over time. Despite the unpredictability of individual turns and movements, he unearthed an overarching pattern on a 3D grid.

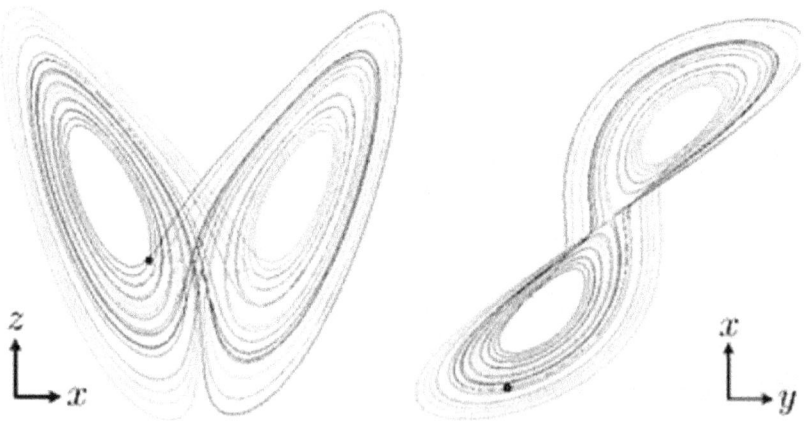

Figure 8.4 Lorenz attractor Graphs by Dan Quinn[7]

Lorenz's discovery of patterns in seemingly chaotic movements is akin to the irrationality of ϕ. The calculation extends beyond our grasp of understanding, yet it gives rise to captivating patterns. While there's a recurrence, the pattern never precisely repeats; it weaves designs that enthrall us in a narrative. Details might seem random and unpredictable, but stepping back to observe the collective movement reveals a pattern. What appears irrational to us often results in delightful symmetry.

There are countless data points and a myriad of systems influencing weather—far more than we can comprehensively track. Consequently, we label it chaos. In reality, it comprises intricate details and interwoven rhythms that elude our tracking capabilities. When faced with incomplete data or a lack of full understanding, we characterize it as randomness, assuming chance is supreme. All the while, it's not sheer chance that hampers predictability for meteorologists. Rather, it is unmeasured and unknown factors to which our weather is exceptionally sensitive. To achieve accuracy, we would have to monitor every data point on Earth, encompassing water evaporation and sun-induced heat from every surface at every conceivable angle. Comprehensively tracking the sun over multiple seasons would be essential. Any long-term prediction relying on data from less than one solar season (approximately eleven years) is scientifically unsound.

History is brimming with discoveries that reveal what we once perceived as chaos, such as the movement of other planets, to be predictable patterns and systems. The symmetry inherent in keeping the universe in motion and sustaining life fosters thriving abundance.

I'm not prepared to embrace total determinism and it is not something math can solve for us. For me, it is an open question. Maybe the chaotic dynamical system is a manifestation of choice embedded in our universe, the parameters of which we have yet to comprehend.

Even less comprehensible to math is beauty. While mathematics can express designs within certain limits, it alone cannot fully capture the experience of beauty. While we can assign a numerical value like ϕ, nothing can substitute the firsthand experience of standing on the Grand Canyon's rim or witnessing orcas playing in Antarctica's crystal-clear waters.

CHAPTER NINE

Life, Meticulously Fine-Tuned

WHEN OUR DEAR FRIENDS, Jessica and her wife Amber, sold their business and embarked on a year of travel, we jumped at the opportunity to explore Ireland with them. The trip was planned to start in Scotland, wind through Ireland and Northern Ireland, and conclude in England, lasting sixty-seven days in total. The entire month before we set out, Trin spent countless hours meticulously planning every detail. He booked twenty-five different accommodations, reserved car rentals, and secured advance tickets to Brú na Bóinne. Flights were researched and purchased. Trin even created a shared map with over three hundred pins, marking the location of each accommodation, must-see sites with detailed descriptions, and parking spots.

Our journey around the Emerald Isle wasn't left to chance—it was a masterclass in planning. Yet, despite all the preparation, the trip still demanded great flexibility and stamina from us all. Typically, Trin and I prefer to travel slowly. Among fellow travelers, we're known as "slowmads"—travelers who linger long enough to soak in a destination's culture and landscape. But with Jessica's and Amber's limited travel window, we opted for a faster pace to make the most of their year off. As it turned out, they probably would have appreciated a little more slomading. By the end of the trip, we were all exhausted—but we had so many fun memories to look back on.

One morning, after an early start, we packed up and got ready to explore before heading to our next lodging three hours north. As we loaded our backpacks into the trunk, Trin messaged the host to let him know we were on our way out. After double-checking the rooms and refrigerator to make sure we had left nothing behind, we climbed into the car, ready for the road ahead.

Just as we started the engine, another car came speeding up the mountain. Since we were at the end of the road, the house we'd just left was the only possible destination for the oncoming vehicle. The car parked next to us, and the host quickly hopped out, walking over to our car.

"Is everything okay?" he asked.

"Yes, everything was wonderful—the views of the lough are stunning," Trin replied.

"Why are you checking out early?" he asked, looking puzzled.

Trin and I exchanged confused glances before he pulled out his booking app. Sure enough, we had made a major oversight—we weren't supposed to check out until the next day!

We thanked the host profusely for catching our mistake and reassured him everything was fine—except for our ability to keep track of dates, apparently. Laughing at ourselves and relieved he caught us, we unloaded the car and reevaluated our plans for the day.

Even with all the meticulous planning, we almost botched the schedule. But without Trin's careful preparation, we wouldn't have experienced the incredible cliff walks, the bike rides across the Aran Islands, or the countless castles we visited. Nearly every day offered breathtaking views, and the meals—whether cooked by Trin or Amber—were nothing short of spectacular.

All this planning pales in comparison to the meticulous fine-tuning required for life on our planet.

Earth revolves around the sun within what scientists often refer to as the Goldilocks Zone, a region where conditions are just right for life to thrive. Positioned at an optimal distance from the sun, Earth maintains liquid water. Were Earth situated closer to the sun, all water would evaporate into steam. If it were farther away, all the water would freeze. Yet, Earth's place within this zone allows for the existence of liquid water, essential for the sustainability of life as we know it. The suitability of Earth's location is but one of the numerous factors contributing to the possibility of life.

More than twenty-five fundamental physical constraints, each requiring precisely the right frequency, finely tune our ecosystem to support life as it exists on our planet. A minuscule alteration, even by a fraction of $1/10^{40}$ in the strength of gravity, would render life as we know it impossible. The current finely tuned fundamental physical constraints are perfect for the formation of stars, the orbiting of planets around them, and the cohesion of all things.

Stephen Hawking wrote in his *Brief History of Time*, "the laws of science, as we know them at present, contain many fundamental numbers, like the size of the electric charge of the electron and the ratio of the masses of the proton and the electron. . . . The remarkable fact is the values of these numbers seem to have been very finely adjusted to make possible the development of life."[1]

Not only is our universe finely tuned, but every element also appears to play a role in the rhythm of life. It is so detailed and co-dependent that we have yet to create a completely self-sustainable biosphere that contains life beyond specific plants.

The Co-Dependent Ecosystem

Small terrariums housing tropical plants, earth, and water in a glass globe are closed systems. Nevertheless, they still rely on sunlight, the energy that sustains life. Gravity is also essential to facilitate moisture recycling and the descent of dead leaves to become soil once more. To flourish, they require meticulous assembly with the appropriate quantity of soil, seed, water, and placement in an environment with the right temperature and optimal sunlight exposure.

Larger biospheres created by scientists supporting organisms that crawl, slither, swim, or walk have yet to achieve success as closed systems.

In 1991, Biosphere 2 was constructed in Arizona, USA, to establish a closed, self-sustaining ecosystem. Eight people were selected to live inside for two years. However, the system failed; it soon ran out of oxygen, most of the invertebrates died, and all the pollinating insects died. Not surprisingly, the populations of the cockroaches boomed. The individuals inside, known as biospherians, reported constant hunger. The experiment contributed to our understanding of ecosystems, but it did not fulfill its initial goal of creating a totally self-sustaining biosphere.

Through greenhouse experimentation, we have discovered trees need wind. Trees grow taller and more rapidly in a biosphere or greenhouse than they do outside, but eventually they topple over. The missing element is wind. Wind compels tree fibers to develop strength and flexibility, slowing their upward growth but enhancing their overall resilience. Without this daily "yoga" of wind, the trees cannot adequately support their weight, causing them to fail in the absence of this crucial element in the pattern.

For a biosphere to function, every component, including oxygen-giving plants and oxygen-breathing beings, would require meticulous balancing. Despite our intelligence and computational capabilities, we cannot recreate the balance of Earth's biosphere. Even with all the intelligence we have accumulated, it still requires more knowledge and understanding of the purpose and balance of each life form. Even if someday we can build a working biosphere, it would still depend on energy from the sun—energy that is far beyond our capability to Create or even copy. Energy input is of paramount importance; without it, all life will die.

Even the space station heavily relies on supplies from Earth. While we might create a terrarium or, eventually, a biosphere capable of sustaining life, the collective intelligence of humanity grapples with the challenge of merely replicating the rhythms of nature. Even with our combined intellect, the intricacies of processes like photosynthesis remain not fully understood.

The Sun: Source of Life

Energy from the sun is essential for life. No life form we have discovered can survive without it. All life forms depend inherently on external energy sources.

Deep-sea creatures, living beyond the sun's reach, subsist on life that descends from the sun-soaked layers above. Bodies of top-dwelling organisms descend into the depths upon their death, transferring sun-derived energy to the extremophiles below. In the deepest, darkest caves, life relies on flood wash to deliver sun-nourished organic material, serving as its sustenance.

The sun imparts the energy of life. It's not something we can purchase or recreate; we freely absorb it. Despite all our advancements, if the sun were to go dark, life would cease.

The rhythm of existence extends far beyond Earth's biosphere—its grandeur spans the universe, reaching beyond what science can measure or humanity can even understand. Its precisely calibrated physical constants cannot even be replicated by mere humanity.

Part 3

The Symphony of Life

Nature does not hide its patterns; they are woven
into the fabric of every life, forming cycles that are impossible to miss.

A truly holistic worldview embraces every element essential to life,
from love to music. Like a
symphony, these patterns unite mathematics and beauty, illustrating
the seamless harmony between science, art, and the human
experience—a rhythm
that embodies love, connection, and the essence
of life itself.

CHAPTER TEN

The Rhythms of Communication in Music

DURING THE MOST CHALLENGING year of my life, I possessed just one music CD, a source of solace. Its lyrics painted a picture of life's beauty, and its calming melodies became a constant companion on my disc player. Jason Mraz's tranquil refrains echoed wherever I went, providing a comforting soundtrack to my journey.

Years after the stress from that challenging period subsided, every time I heard Jason Mraz singing about how wonderful life was, anxiety washed over me like an unbidden memory. What had once been a source of comfort during that tumultuous year had become entangled with the stress of that time.

Music transcends language, weaving itself into the fabric of our memories. It doesn't merely resonate with our minds; it engages our entire brain and body. Scientific studies on brain development reveal an expansion of the corpus callosum when learning a musical instrument, indicating enhanced cognition not only in the right hemisphere but throughout the entire brain. Music serves as a profound communicator resonating with our entire being in ways we cannot explain.

Beyond the cognitive, music speaks to our emotions, urging us to sway to its rhythm. While there's an ongoing debate about the superiority of digital versus analog music recording, neither can rival the experience of a symphony hall. A live orchestra produces a richness and depth that cannot be replicated in a recording; something is lost in the copy.

Music serves as a bridge between mathematics and the softer sciences of emotion and movement that defy simple formulas. Although there is an entire science devoted to studying music and its interaction with various brain regions, the emotions music communicates and evokes remain elusive to concrete scientific descriptions.

Using verbal communication, patterns, and rhythmic interplay, music weaves these elements into a tapestry that tells a rich story capable of evoking joy or bringing tears.

Music is both cultural and experiential. Listening to Jason Mraz sing "Life Is Wonderful" is unlikely to cause anxiety for many people. My anxiety stemmed from the specific experience I had while listening to it. Music has the power to transport us back to moments in our lives, playing on our emotions. It can stir a crowd or provide solace to an individual. Even before children can speak, they instinctively sway to the rhythmic beats.

As I composed this chapter, I found myself in an apartment in Rabat, Morocco. The night before, our neighbor had passed away, and this morning, the anguished cries of those left behind reverberated throughout the neighborhood. The sorrow was palpable. No words or numbers could soothe this sorrow. I didn't know the meaning of the Arabic words they were using to sing their mourning songs, and I didn't even know the name of the departed soul. Yet, I sensed the sorrow transcending language, reaching us through their heartfelt melodies. Singers had harmonized a cappella, creating poignant tunes that echoed through the night. It was a sound of grief resonating through the streets, somehow providing solace. Music, even when we couldn't decipher every emotional-chemical interaction, stood as a potent communicator. It existed beyond the confines of scientific description.

Music is powerful in more ways than we can describe. It even seems to help with healing. "Biomedical researchers have found that music is a highly structured auditory language involving complex perception, cognition, and motor control in the brain, and thus it can effectively be used to retrain and reeducate the injured brain."[1]

Beyond mere lyrics, music serves as a communication bridge that connects cognition, emotion, and instinct.

Paleontologist, evolutionary biologist, and historian of science Stephen Jay Gould believes music is merely a byproduct of human brains that have grown large during the process of evolution. In 1982, he introduced the term "exaptation" to describe traits whose function is distinct from the reasons for their historical origin. Gould proposed music is a skill not directly selected for and lacks a specific purpose in the process of evolution.

Music transcends our current scientific understanding and doesn't neatly align with Gould's theory of origins. Instead of questioning the theory that music doesn't seem to fit into, Gould categorizes the unexplained as an exaptation. Because of his steadfast commitment to an origin theory, Gould dismisses any intrinsic purpose in music, hindering the search for alternative explanations.

The term exaptation appears to be a contemporary twist for an age-old concept known as "vestiges." In 1893, anatomist Robert Wiedersheim published *The Structure of Man, an Index to his Past History*,[2] which included a now-obsolete list of eighty-six organs or body parts presumed to be remnants from a previous evolutionary stage with no significant purpose in contemporary human bodies. The list encompassed organs like the thymus and pituitary glands, as well as white blood cells and other various vital functions, all deemed to have no known function in 1893 and therefore labeled as evolutionary leftovers.

We now recognize the crucial role these vestiges play in maintaining our health and equilibrium. "Junk DNA" represents another contemporary manifestation of the age-old vestiges notion. Molecular biologists, however, who believe there is purpose and intelligence in the design of life, investigated further and discovered this non-protein coding DNA—labeled as junk by some—directs other processes.[3]

While vestiges and exaptations differ—one refers to a trait that has lost its original function, while the other describes a trait repurposed for a new use—neither serves its initial purpose. One has theoretically been adapted for something new, while the other has diminished or disappeared. But if past assumptions about vestigial structures were wrong, could exaptations be similarly misunderstood? Neither has proof that it was used for any other purpose.

Some evolutionists accuse intelligent design theorists of holding a "cop-out" explanation, yet couldn't we label exaptations, vestiges, and junk DNA the same? Intelligent design is a pattern we can observe. Design encourages us to understand the how and why of life—it promotes scientific research. Vestiges, however, end the search. Labeling something as junk or left over literally concludes there is no purpose or meaning to something, so why investigate further?

We repeatedly classify what we don't comprehend as chaos or remnants, only to realize later they serve a specific and vital function in the interdependent rhythm of life. Our discomfort with saying, "I don't know" leads us to offer erroneous explanations for the existence of an organ or an unpredicted outcome. Surely, we tell ourselves, it can't be our limited knowledge; it must be the world is chaotic. Scientists are continually uncovering new facets of the human body, yet much remains beyond the scope of precise mathematical understanding. Believing there is a design fuels curiosity.

The founders of the enlightenment, such as Isaac Newton, Galileo Galilei, John Locke, René Descartes, Gottfried Wilhelm Leibniz, Voltaire, Rousseau, Montesquieu, Immanuel Kant, and many more, were theists or deists. They studied because they believed there was a design to be found.

The Elixir of Music

Music possesses the capacity to evoke romance, patriotism, motivation, or contemplation. In various cultural perspectives, music is intrinsic to worship. Even those who don't believe in a god harness music to worship what they hold dear, including love and sex.

Music communicates love and emotion, integral aspects of the human experience that often imbue life with meaning. It elevates a singular fact into an immersive experience, engaging us beyond our logical minds.

Some evolutionists term music an exaptation because of its elusive nature beyond complete scientific formulation. Creationists consider music a gift of communication. Even though we cannot prove why music exists or how it originated, its vital role in expressing the ineffable is undeniable.

A friend of mine was rehearsing a song she planned to perform. Her entire focus was on attaining perfect pitch and precision on each note. Her hyper-focus on exactness created a performance that seemed void of emotion. Mathematically, her rendition was flawless, yet it was challenging to listen to. The intense emphasis on accuracy failed to convey the emotional essence the song intended to communicate.

In the realm of science (to my knowledge), there is no evidence to prove how a song transmits emotion, but we unmistakably sense its presence. It adds an extra layer of powerful communication. Despite my fondness for math, life's richness extends beyond equations, and abundance is beyond the spreadsheet.

CHAPTER ELEVEN

The Rhythm of Pair Bonding

SCIENCE, WITH ITS FORMIDABLE ability to reveal the intricacies of the physical world, has yet to unravel the mysteries of passion, joy, or love. Science shows us how things work and how they interact, and it provides us with evidence of connectivity and patterns. But true freedom of the heart, mind, and soul eludes formulas. It's akin to the wind—evident in its effects, though invisible to the eye.

Love defies reduction to a mere checklist. When seeking love and acceptance in the gaze of others, it is impossible to define it mathematically.

Words extend beyond the confines of mathematical limitations, and emotion exceeds them both. Even without a shared language, one can perceive the sentiment of another person. In foreign lands, despite linguistic differences, I've communicated my love for the delicious creations of food to vendors. They understood and responded with beaming smiles, reflecting joy in their accomplished work. Emotion and nonverbal communication bridge the gap that words of varying languages create. Emotion operates beyond the limits of science.

Our minds, craving precision and control, often find the realm of emotion challenging. In Patti Callahan's book *Once Upon a Wardrobe*, a poignant exchange unfolds between a mother and her daughter. They stand in the kitchen, grappling with the impending death of George, the youngest son and sibling. The mother confesses, "I don't know the right answer to anything these days." Her daughter responds, "Neither do I, Mum. I don't know if anyone does. Only math problems seem to have right and wrong answers, far as I can tell lately."[1] The mother says nothing, holds her daughter's hand for a moment, then turns to make dinner; her response is more of an answer than words can express. She does what she must, even amid uncertainty.

Life is full of unknowns, but the unknown sparks our imagination and gives us joy in discovery. The unknown can also bring anxiety and fear, prompting a quest for certainty. We like to have answers, and so, as a culture, we have turned to science for all the answers, attempting to describe love through chemicals and brain reactions. But these explanations fall short of capturing the essence of life.

Pragmatic and formulaic subjects offer comfort, but to embrace the aspects of life that make it worthwhile, we must step into the unfamiliar and venture beyond the confines of mathematical limits.

Steve Wozniak, a key figure behind the technology that now provides us with Apple phones, excels in the realm of science. Yet, while he mastered the complexities of circuitry and innovation, navigating social dynamics proved far more challenging. In Walter Isaacson's biography of Steve Jobs, there's an excerpt that highlights this contrast:

Isaacson said, "Woz [Stephen Gary Wozniak] became more of a loner, boys his age began going out with girls and partying. Endeavors that he found far more complex than designing circuits."[2]

Arguably one of the most important aspects of thriving is love, yet it is difficult to find funding for such a subject as neuroscientist Dr. Stephanie Cacioppa discovered. She faced challenges when attempting to study the effects of love on the brain. Reviewers summarily dismissed her initial research proposal which focused on love. Then she tried submitting the same proposal, replacing only one word. The new proposal received approval and funding. She changed "love" to "pair bonding," underscoring how love remains too elusive for precise scientific studies.

Humanity has long wrestled with the concept of love. In the thirteenth century, Emperor Frederick II allegedly experimented with children, forbidding their caregivers from speaking to the babies in their care. Frederick wanted to see what language they would develop once their vocal cords matured. Unfortunately, the experiment also deprived the children of touch, leading to their failure to thrive and eventual death. His diabolical experiment highlighted the crucial role of love, or at least the affection of touch.[3]

In her book *Wired for Love*, Dr. Stephanie Cacioppo presents evidence that love is a biological necessity. Cacioppo's neurological study further supports the notion that humans require love—both receiving and giving it. She explains how love activates specific neural pathways, particularly in the prefrontal cortex, which are essential for motivation, reward, and emotional regulation. Neurotransmitters like oxytocin and dopamine play a crucial role in bonding and attachment. Her research shows romantic love is not only a human need but also essential for our overall health.[4]

Fractured Love

There is a discernible fracture in our cultural fabric and a growing number of people who are giving up on love. According to Pew Research, a staggering half of single adults and a majority of single women assert they are not currently in the dating market.[5] In America, 61% of millennials are living without a partner.[6]

Living independently does not equate to loneliness or a lack of love. Even Dr. Cacioppo, before meeting the love of her life, reported not feeling lonely. She remained connected to friends and family, surrounded by love. Still, *Psychological Science* reports a rising trend of individualism.[7]

In 2017, former U.S. Surgeon General Vivek Murthy labeled loneliness as a public health "epidemic." Following suit, a year later, the United Kingdom appointed a "minister for loneliness."[8]

Humanism, one of the dominant belief systems of our time, can be traced back as early as 490 BCE with the ideas of philosophers like Protagoras, who emphasized human agency. His famous statement, "*Man is the measure of all things,*"[9] reflects a human-centric view of the universe. Modern advocates of this belief promote a self-centric philosophy, persuading individuals they do not need others. Independence is celebrated, and those who "do it all themselves" are admired. Even in films, the trend leans toward depicting characters who end up alone, framing it as a heroic act of strength to forgo relationships. Greta Gerwig's 2019 adaptation of *Little Women* is a prime example of subverting the traditional romance narrative, where the protagonist, Jo March, chooses her career over pursuing a relationship.

Yet human responses often swing like a pendulum. This shift in choosing to be alone may be a reaction to a history in which women were frequently treated as mere objects. But taken to an extreme, it risks discarding the deep intimacy and connection that companionship offers—connections. Nature continuously illustrates the need for connection in its intricate patterns.

Individuals, with or without partners, can cultivate fulfilling relationships. But opting for singlehood is not an act of heroism, just as engaging in a relationship is not a sign of weakness.

Individualism with its emphasis on self-knowledge has once again been elevated to a cultural ideal. The roots of this belief in reason and autonomy can be traced back to the philosophical teachings of Socrates, Plato, and Aristotle. However, theologians might argue that its origins stretch back even further—to the first humans who sought self-knowledge by reaching for the forbidden fruit, prioritizing it over their connection to God. This cultural trend is reinforced by the common adage, "If you don't look after yourself first, no one else will." It mirrors the old geocentric view, where we once believed ourselves to be the center of the universe, in full control of our destiny. Yet, the pursuit of control—especially over things or people beyond our grasp—creates a dissonant undertone within our culture. The Healthy Marriage Organization identifies control issues as a leading cause of divorce,[10] underscoring that control, when wielded in relationships, is ultimately detrimental.

Intimate relationships are a delicate dance, requiring adherence to the patterns that bind our universe together. We may have learned in the sixteenth century that the physical universe does not revolve around us, but we persist in placing ourselves at the philosophical center. Instead of nurturing connections and being part of a community, we stand on the pedestal of individualism, erroneously believing we can navigate everything alone.

Harvard Health highlighted the detrimental health impacts of loneliness and equated it to smoking a pack of cigarettes a day.[11] This emphasizes the fundamental human need for love.

Isolated in a Sea of Knowledge

We find ourselves in the era of data, where the wealth of human knowledge has expanded exponentially.

Words and their combinations construct an edifice akin to the Tower of Babel.[12] In pursuit of knowledge with vast databases and quantum computers, some believe we will create heaven on Earth—a utopia. Yet, amid the advances in knowledge, we seem increasingly ensnared in distractions—not just through our devices, but by the constant noise and advertisements that pull us away from what is truly fulfilling.

Ironically, the very words we use to communicate are pushing us further apart. Even a sentence spoken in kindness can lead to someone being "canceled" if that individual unknowingly uses a word that once held no ill intent but is now deemed offensive. As we continue to redefine words, disregard context or the speaker's intent, and take offense at the slightest provocation, our ability to genuinely connect becomes increasingly strained.

As data becomes more readily accessible, we seem to lose the context from which it came. Artificial Intelligence is like a search engine on steroids. I love how quickly I can find some answers. However, I miss the opportunity to learn many other things along my quest for that answer. Now, instead of reading an entire book or at least an encyclopedia article, I can just ask AI when a specific event occurred.

Much like the positive and negative charges inherent in every atom, every technological innovation carries both virtues and drawbacks. With our modern technology, the downside manifests as increasing isolation. Our homes, equipped with double-pane windows to block out external noise and regulate temperature, insulate us from the outside world. In our effort to escape the traffic and noise pollution we've created, we also silence the natural sounds that once connected us to our environment.

While we stay informed about events around the world, it's easy to overlook the neighbor just a few feet away. Our daily routines make it simple to distance ourselves from the impact of our actions—we flush away waste without a second thought, leave trash at the curb for someone else to handle, and move between climate-controlled spaces, often unaware of the natural cycles that sustain us.

Our cities, secondary creations, increasingly insulate us from the natural world. Sadly, conversations often devolve into written words, brief texts, or mere emojis. We find ourselves increasingly alienated from both each other and nature. How much time does most of today's civilization spend in contact with or observing the original Creation—anything not crafted by humans? How often do we breathe fresh air outside?

Charles Eisenstein, an American author and activist, observed, "not only is technology based on a conceptual separation from nature, but it also reinforces that separation. Technology distances us from nature and insulates us from her rhythms."[13]

We profess to rely on science, but neglect the repeated patterns and connections that make life thrive. Despite evidence of the detrimental impact of breaking connections, we plow forward, thinking we can do it all alone.

The Price of Comfort

When traveling, Trin and I rarely stay in hotels. They are the same everywhere and lack the interaction with local life that other types of accommodation can provide. On one particular evening in England, we arrived at the home where we would stay and met Mary, the homeowner and our host. She warmly welcomed us inside. I liked Mary right away and was curious to know more about her.

That evening, I bought a bottle of wine to share with Mary. Queen Elizabeth had passed away the day prior, and the television played a continuous broadcast of programs concerning the royal family. However, the television news and photos of the royal family faded into the background as I listened to her stories.

Mary was born into abject poverty. Her family lived in the slums with little more than shelter from the rain. Meals were prepared next to neighbors in open-air kitchens. With shared latrines and outdoor kitchens, interactions were unavoidable, and there was scant privacy.

Eventually, the government intervened and provided low-income housing equipped with kitchens and indoor plumbing. Reflecting on this transition, Mary remarked, "We moved into a house, and we lost the community and friendships we had." While grateful for the improvement in living conditions, she also conveyed the loneliness that accompanied it.

Mary didn't sugarcoat the desperation of life in the slum, but in every stage of her vibrant life, she extracted the positive aspects. Despite the challenges, she cherished the bonds of friendship forged amidst extreme poverty. The sense of community brought joy even in the face of adversity. The privacy afforded by their new home felt isolating to her.

Yet, outsiders reflecting on historical hardships often focus on conditions they have not experienced and fail to consider the nuances and positive experiences that Mary understood.

I strongly support the importance of housing and I value my privacy. But with each comfort we attain, we inadvertently shift closer toward isolation. Finding balance is crucial, and I don't wish poverty upon anyone. Mary's perspective certainly prompted me to contemplate our cultural drift toward isolation amidst increasing comforts.

Having spent years on the road, Trin and I often lodged in rooms within others' homes. Despite occasional discomfort and a lack of privacy, this experience has blessed us with countless beautiful memories and friendships around the world that would have been unattainable in a private dwelling.

Returning to the USA for visits always required an adjustment back to hermetically sealed homes. In South America, people left windows open regardless of the weather. Many windows wouldn't close, leaving gaps that allowed a breeze to flow through the house.

During our time in Australia, we lived in a converted Toyota Coaster without air conditioning, opting to keep the windows open instead. It felt as though we were living outdoors, albeit with the convenience of a roof, kitchen, shower, and toilet. Reflecting on this, I'm grateful we didn't opt for air conditioning. Embracing closeness to the world outside is one of the most memorable and epic aspects of our travels.

We often view our gadgets and luxurious homes as symbols of progress, but we are inarguably increasing our disconnection from each other, the world outside, and even our own personal conflicts. Do we destroy something in another part of the world so our own personal comfort can increase? Somehow, humanity misuses every good invention. Rockets let us go to space but are also used to launch missiles. Our disconnection from each other is evident in our warfare. Battles used to be fought hand to hand, or at the very least, with the enemy in sight. Now, humankind can engage in mass killing without ever confronting our opponents face-to-face, just by pushing a button. Is this progress?

The patterns of nature show us a co-dependent ecological system. Every plant and animal, wind pattern, and raindrop, provides a note that plays its part in the symphony of abundant life. Yet, in our pursuit of comfort, we trade those connections, ignoring the detrimental impacts.

CHAPTER TWELVE

The Narrow View of Specialization

THE PATTERNS OF NATURE explored so far in this book are mere glimpses of the numerous captivating patterns that orchestrate the rhythms of the universe we inhabit. Each field of study is so profound it could fill libraries with data and still fall short of capturing its complete essence. Every life form and object in the solar system exhibits fine-tuning and interdependence. To comprehend the grand scheme, we must recognize how it all harmoniously aligns, similar to Lorenz's unpredictable waterwheel generating overall patterns. Physicists seek a theory of everything—a singular theory unifying our universe, a consilience of patterns.

In 1998, Edward Wilson, a pioneering American biologist and naturalist, best known for founding the field of sociobiology, penned *Consilience: The Unity of Knowledge*, delving into the history of the growing body of knowledge shaping our modern world. In antiquity, when written knowledge was not readily available, philosophers—the scientists of that age—mastered an array of subjects. In today's world, where vast databases put endless information at our fingertips, diving deep into any single subject is challenging, as the sheer volume leaves little time for broader exploration. Wilson contends, as a result, we risk losing the overarching consilience or amalgamation of knowledge. He argues that relevant solutions emerge from a broad understanding of the sciences combined with cultural studies. Only then can we grapple with the elusive aspects of human existence, like love, that resist easy categorization or formulation. Wilson emphasizes one cannot adequately address social issues in isolation; a holistic approach that considers all factors is imperative.[1] A wider breadth of experience and study helps us see the overall pattern, like Lorenz did when he stepped back to watch the pattern of the waterwheel.

The Narrow Boat

We first met James as he and his partner, Kelly, were sailing their narrow boat up the Thames. After watching them deftly operate the lock, we were invited aboard to sail with them to the other end of Oxford. During the journey, we discovered that both James and Kelly are incredibly accomplished, with brilliant careers, yet they cherish the tranquility of life on a boat. To some, this might seem like a contradiction, but to me, it felt like a harmonious balance—their professional successes complementing the serenity of their lifestyle.

James worked as a consultant, advising CEOs on critical decision-making. He emphasized the importance of "T knowledge"—a blend of deep, specialized expertise in a specific field and a broad, general understanding of the larger context. This concept resonated with my experience in corporate information technology, where professionals with diverse experiences often excel at solving complex problems. They see challenges from multiple perspectives, drawing on patterns from various disciplines to craft innovative solutions. Patterns, after all, repeat, and recognizing them can be more valuable than a college degree.

James and Kelly embodied this philosophy, navigating the patterns of the river and the peace it offers, while sailing through the demands of the business world.

In our race toward specialization, we risk losing knowledge of the interconnectedness of the entire ecological system. The massive body of knowledge leaves little energy for individuals to explore disciplines outside their area of focus. While a mycologist can captivate us with insights into fungi, the same expert may have limited knowledge of the sun's patterns, seasons, storms, and cycles. It is unreasonable to think that one person could be an expert in every discipline. The result of the grand body of knowledge is that pursuing expertise in one discipline often comes at the cost of the bigger picture.

Edward Wilson highlights this dilemma by emphasizing the essence of a scientific career lies in pursuing discovery. Scientists are defined by their capacity to uncover new knowledge and make groundbreaking findings that advance our understanding of the world. Discovery is paramount if one is to be considered a good scientist.[2]

Wilson underscored the issue, stating, "they learn what they need to know, often remaining poorly informed about the rest of the world, including most of science for that matter, in order to move speedily to some part of the frontier of science where discoveries are made. . . . The difference explains why so many accomplished scientists are narrow, foolish people, and why so many wise scholars in the field are considered weak scientists."[3]

Our cultural emphasis on individualism—a myopic focus on self—and our definition of success not only shape our private and corporate world but also influence the scientific community. Discovery is synonymous with the pursuit of scientific success, yet this pursuit often compels scientists to immerse themselves so deeply in their field that they sacrifice a holistic understanding of interconnected patterns across disciplines.

There is a vicious competition within scientific fields of study.

"[Funding] affects what we study, what we publish, the risks we (frequently don't) take," explains Gary Bennett, a neuroscientist at Duke University. It "nudges us to emphasize safe, predictable (read: fundable) science."[4]

The overlooked insight from Noether's discovery—that every constant symmetry leads to higher truths—underlines the necessity of incorporating the similarities and interactions across all fields for a comprehensive worldview. The big picture, the symphony of life, reveals the Truth that holds everything together.

When Specialization Harms Health

Industrial monoculture—single-crop or single-species farms, often mechanized—starkly illustrates a disregard for consilience and the natural rhythm. This approach, driven by the pursuit of increased production, neglects interactions with the broader natural world, akin to trying to roll a cart with part of a wheel missing. The resulting jolts and rattling from the misshapen wheel will eventually break the cart into pieces, just as our industrial monoculture production creates a domino effect of destruction.

These farms confine one species into inadequate spaces. Resorting to various chemicals and injections that sterilize everything, they inadvertently foster superbugs. Injecting hormones to boost the mass of the animals further illustrates their focus on quantity over quality. The animals, the health of those consuming the animals, and the land we scorch are paying the price. Cramped conditions create unhealthy environments, turning food sources into breeding grounds for bacteria and subjecting animals to misery.

Harvard Public Health Magazine states that "about 678,000 Americans die each year from chronic food illness. That toll is higher than all our combat deaths in every war in American history—combined."[5]

Conversely, the total number of deaths from COVID in 2020, according to the CDC, was 350,831.[6] COVID was a health concern we were right to take seriously. Should we not also take the deteriorating state of our food system seriously? We're so focused on providing inexpensive food that the poor quality ends up costing us more in healthcare than we save.

A few years ago, I experienced heavy heartbeats. When they started making me dizzy, I immediately scheduled an appointment with my doctor. I told the doctor I had recently gone on a new diet and was participating in an advanced spin class, along with swimming miles every week.

She took tests and bloodwork. When the bloodwork came back, her nurse called me and said I needed to take Lipitor immediately. I reminded her I had eaten just before my blood test and was told they could not test for cholesterol.

"Oh, okay, ignore this then," the nurse told me. I respect doctors; they are knowledgeable and seek to help their patients. Despite their high level of education, I don't consider them as infallible beings who should never be questioned.

Doctors often seek mathematical certainties in diagnosing complex biological systems, leading to potentially erroneous conclusions. I ignored the ill advice and changed my diet—the diet she never asked about, even though I provided the lead-in. My symptoms went away. Specialization is necessary, but without knowledge of how specialized fields interact with everything around it is dangerous.

My story is not unique. I hear the same from friends and family. They see specialist after specialist, none of them talking to each other or looking at the entire picture. They undergo testing to find mathematical certainties, and no one can figure out what is going wrong because the body is not a machine. The body is an amazingly complex organism that still eludes our ability to map its functions and interactions completely. In my case, the misdiagnosis and directive to just take a pill resulted from the loss of a holistic approach. I am not alone in this erroneous diagnosis. A study published in the BMJ Journal estimates 795,000 Americans die or become permanently disabled because of misdiagnosis each year.[7]

I'm thankful for the progress we have made in medicine and so glad we no longer use leeches for bloodletting. I am grateful for Novocain and sleeping agents that eliminate the horror of surgeries. We can conduct surgery that saves lives and give medicine that heals deadly infections. But, in our rush to specialize, we are losing the interactions of the bigger picture and thus starting to lose the game.

I understand the threat of lawsuits might make doctors want proof from a test before a diagnosis. Then they seek some pharmaceutical company-approved cure (I'm not even going into the pharmaceutical issue here). Our bodies are complex, and there are so many lifestyle changes that can improve our health. But so many would rather prescribe a pill or take a pill—a quick fix. Quick fixes are a mechanistic solution that ignore the complex biology of life. Sometimes medication is necessary; many times we just need to rebalance the rhythm of life. Diet, exercise, stress, and personal connections all play a role in our health.

If we rely only on what we can mathematically prove, we have missed the overall picture and most likely misunderstand even what we are myopically studying.

The U.S. healthcare system, among the most expensive globally, ranks 48th in life expectancy.[8] We have amazing doctors, and I trust the hospitals in the USA more than in many other countries, but that doesn't mean they are perfect. Finding and admitting where we are breaking patterns will help us on the path to freedom from many of the diseases we have created with our broken food production.

We have made great strides in the survival rate of infants that are born and vaccinated against many diseases, but we have increased the prevalence of diabetes, cancer, and heart disease.

Disconnection causes not only health and healthcare issues but also, as mentioned before, increases stress and loneliness.

The Big Picture

The entire ecological system unfolds in breathtaking complexity as we observe the patterns of each species and genus, interacting through nutrient cycles and complex relationships. A doctor must understand not only the physical body and the chemical reactions of our hormones but also the microbiomes that live in and on us. Environmental factors such as stress, pollution, diet (including the diet of the animals we consume), allergens, and more play a crucial role in our overall well-being.

Beyond individual health, the stewardship of our planet requires the study of everything.

Our actions show how disconnected we are from our neighbors and environment. For example, releasing balloons to signify the liberation of sorrows may provide personal solace, but it overlooks the environmental impact of litter. While staying in Western Australia, a neighbor in Burkup County shared a heartbreaking story with us. Her beloved cow was bleating in distress, mourning over her lifeless calf, which lay motionless on the ground. The calf had suffocated, a discarded balloon wrapped around its mouth and nose. What symbolized the release of sorrow instead became an instrument of death, bringing grief to the mother.

All our actions impact someone or something else. Litter is my pet peeve, especially when it is the result of an over-consumer producing loads of trash to ship off to a poor country. As we travel, I see plastic everywhere—in the oceans, across the fields, and over the countryside— in the lands of those who cannot afford to collect it all and ship it to someone else. But we, disconnected from our neighbors, take little notice, as long as plastic is convenient for us.

Consilience across various sciences, from astronomy to agriculture, is crucial for informed decision-making and effective policy formulation. Instilling fear and enacting ineffective policies that serve only to check our virtue-signaling boxes are detrimental. Understanding the patterns and initial conditions influencing climate, the seasons of the sun, solar storms, the earth's tilt, wind patterns, volcanic activities, scorched-earth farming, and other factors labeled as chaotic by Edward Lorenz is essential for grasping the complexities of our world—a symphony of interconnected patterns.

With all of our scientific advancements and accumulated knowledge, why does humanity continue to chase after broken rhythms? Perhaps chaos theory deserves more credit than we give it. The initial conditions of our belief systems profoundly shape the trajectories we follow. Let's take a step back from beautiful detailed patterns and explore how our belief in origins guides our understanding of ecological systems and humanity itself.

Part 4

Patterns, Purpose, and the Blueprint of Life

Beliefs and experiences shape how we perceive and interact with the world. While personal beliefs don't alter the fundamental patterns of life, they do influence how we interpret and respond to them.

Aligning with the universe's rhythm requires a thoughtful reexamination of our foundational beliefs about humanity, purpose, and origin.

Faith acts as the "initial condition" that sets the trajectory of our lives, connecting us to life's rhythm—or creating barriers to it.

The patterns of life and nature hold answers.

CHAPTER THIRTEEN

Miracles—Breaking Patterns

SITTING IN THE COZY front room of Erratic Rock,[1] a combination hostel and gear rental shop in Puerto Natales, Chile, we listened to Rustyn, the co-owner, detail what we might expect over the next few days. He painted a vivid picture of driving rains, bone-chilling cold, and fierce winds that could blow us off the trails.

"The trails are well marked in Torres del Paine; you can't possibly get lost," Rustyn assured us, taking a sip from his Mason jar. "Yet," he added with a hint of seriousness, "every year, someone gets lost. Things happen. Be smart, stay on the trail, and be prepared for any weather."

The allure of Torres del Paine National Park lies in its breathtaking landscapes, which are among the best in the world. Located in Patagonia, Chile, it boasts a terrain of jagged mountain ridges, glaciers clinging to the rugged landscape, and streams that cascade into turquoise lakes. These waters sometimes mirror the sky with a tranquil surface, while at other times their breaking waves hint at the land's ever-changing temperament and add a layer of drama to the already spectacular scenery.

Trin and I were on the cusp of beginning a multi-day trek, equipped with our tent, camping gear, and food. On the morning of our departure from Puerto Natales, we hoisted our bags and headed for the bus that would transport us into the heart of the National Park. Despite layering up, a chill seeped into our bones, excitement mingling with a tinge of apprehension. Our aim was to immerse ourselves in the park's wonders and ultimately ascend to the iconic towers of Torres del Paine.

On our first full day, we hiked up to Gray Glacier. The day unfolded beautifully, albeit interspersed with bouts of hurricane-force winds. Thankfully, we stayed dry; getting drenched and then chilled to the core was one of my biggest concerns for this trek.

The following day we planned to hike up the French Valley. Rustyn warned us that frigid winds, cooled by passing over the glaciers in the west, pick up speed as they squeeze through this narrowing valley. He also warned us of the sideways, driving rain that shoots out of this valley. We packed our gear, ready for the worst, and headed toward it.

Contrary to our expectations, the sky was clear, and a gentle breeze accompanied us. The vistas of Valle Frances and Mirador Britanico were breathtaking, arguably rivaling even the towers themselves.

A particular mountain captured my attention, standing boldly against the azure sky. Trin and I affectionately dubbed it "Cupcake Mountain" because its white sides and dark, dome-like summit resembled a chocolate cupcake's top, overfilled and cracked. By this point, the trail mix had clearly worn on us.

Occasionally, we heard the thunder of Glaciar Francés calving. We would stare at the mountain and find a new waterfall caused by the ice breaking away and releasing a pool of water from above. The waterfalls were only momentary, soon trickling to a stop.

That night, snug in our tent, we were audibly reminded of the French glacier's might. The sounds of ice cracking and tumbling down rugged cliffs were akin to thunder, drawing us closer to the raw power of the park and its icy summits.

The forecast for the following day was disheartening: 100% cloud cover, which did not bode well for our hopes of witnessing the towers at the peak of our hike on the last day. Yet, knowing the park's weather could swiftly change, we had to give it a shot anyway. I sent up a prayer, asking God to show us the beauty of His Creation that day.

Dressed in several layers, we began our trek before the sun rose, hitting the trail by 5 a.m. A biting wind from the valley met us head-on. I secured my wool scarf and zipped up my down coat, ready for the 20-kilometers that lay ahead and a little over 2,400 feet of elevation climb.

As we climbed into Ascencio Valley, the icy wind offered a respite to the warmth we generated on the incline trail. The dance of putting on and taking off my coat became routine, a strategy to stay dry and avoid a deeper chill that could set in from the dampness of sweat.

The wind intensified, forcing us to lean into it to keep from being knocked down. This was the notorious wind of Torres del Paine, capable of sweeping tents away—even with people in them. We carefully kept our distance from the trail's edge, mindful of the adjacent ravine.

We appeared to be skirting the edge of a storm. The darkness of rain shrouded the valley ahead, occasionally misting us before the wind swiftly dried the droplets away. It moved forward, keeping with our pace.

A vivid rainbow formed in the valley, growing in intensity as we neared, painting one of the brightest spectrums I'd ever seen. It gave us hope, and we pressed on above the tree line, navigating through giant boulders to the hidden glacial lake at the trail's end. Clouds veiled the towers' peaks, yet the sight was still breathtaking. A handful of others shared the moment, hopeful for a clearer view.

Settling on a rock, we layered up against the cooling air, watching as the strong winds blew mist over the water surface, creating fleeting rainbows that danced in the few rays of the sun peeking through.

For the next half hour, the clouds on top of the towers morphed and undulated in a slow whirl that gave us views of one tower at a time. We continued to wait until, at last, a wind blew in and cleared the clouds, giving us a full view of the towers with a blue sky behind them. We stood there in awe, not just for the majestic view but also for how special it felt, after all the prep and anticipation and the hiking for the past few days to get there.

An hour passed, and clouds once again descended upon the towers and the cold became uncomfortable. We took one last look and began the return hike.

Excellent weather and stunning views blessed our visit to Torres del Paine, a place of breathtaking majesty. Some might call the clear skies and the dramatic unveiling of the towers a miracle. Scientifically, however, such events are entirely plausible, with probabilities that fall well within reason for an optimistic forecast. The frequent winds of the park suggest the revealing of the towers might be a routine, natural occurrence.

Yet, to us, it felt spectacular—it felt like a blessing. While what we experienced didn't break any laws of science and therefore doesn't qualify as a miracle, that doesn't mean God couldn't have used the natural wind He Created to bless us. Others might attribute our experience to mere chance. Proving either scenario is impossible, because interpretation lies in the observer's eye and belief system.

Numerous stories tell of people seeking divine help in desperate times, like the man who got lost in the Alaskan wilderness during a storm. Believing he faced certain death, he prayed fervently, promising to believe in God if he survived. Later, while recounting the event to a friend in a bar, someone asked, "You're alive—do you believe in God now?"

"No," he replied. "An Inuit man came along and saved me."

Perhaps it was merely a local man with a deep historical knowledge of the area and weather. Or perhaps a divine power sent that man. Either way, it wasn't a miracle, as nothing broke the laws of nature. The distinction between divine intervention and coincidence is a matter of perspective, shaped by faith—in chance or in God. While people might casually refer to such instances as miracles, this book focuses on a more precise definition of the term. We will leave discussions of divine intervention for another book.

What Is a Miracle?

David Hume, an influential figure in the eighteenth century, primarily receives recognition for his system of philosophical empiricism, skepticism, and Naturalism. He defined a miracle as a "violation of a law of nature."[2]

C.S. Lewis further elaborates on Hume's ideas, stating, "The more often a thing has been known to happen, the more probable it is that it should happen again, and the less often, the less probable.... A miracle is therefore the most improbable of all events."[3]

In this book, the term "miracle" refers specifically to events that explicitly and unmistakably violate the laws of nature—the consistent patterns our universe follows. Observing a face in a rock formation or seeing shapes in the clouds might feel meaningful, but these are not miracles. Similarly, recoveries from cancer that are considered medically improbable, though extraordinary, often remain within the framework of natural biological processes and do not break the established patterns of organic life. Some healings are so extraordinary some may deem them as divine intervention, yet they still reside within the realm of possibility (they are scientifically possible), even if the probability is small. For clarity, a miracle involves a clear and undeniable disruption of natural laws—where the predictable patterns of science are violated.

For instance, a person declared dead, showing clear signs of physical death such as blood separating into water and plasma, only to resurrect three days later, would constitute a miracle. Likewise, the genesis of life from non-living matter, or abiogenesis, defies all observable biological patterns and scientific laws.

In science, the events that violate natural laws are called singularities. Both a miracle and singularity are defined as events in which established laws of physics break down, and known mathematical models no longer apply. These phenomena defy the current observable and/or experimental functions or laws of our natural world.

The key distinction between a singularity and a miracle lies in the perceived cause. Singularities are attributed to chance or an unknown force, while miracles are typically linked to the intervention of a deity or supernatural power. Though often used interchangeably, their implications differ.

Theories of origin, whether framed scientifically or theologically, inherently rely on miraculous steps. Cause and effect is a basic law of science. There are miracles in every origin theory—some attribute the cause to an intelligent being, some attribute cause to unknown forces or mere chance.

Believing in either a miracle or a singularity ultimately requires faith.

Faith and Patterns

Faith shapes the way we interpret the patterns around us. At its core, it is built on our beliefs about origin. Is the universe guided by order and purpose, or is it a product of randomness and chaos? Can we recognize patterns and learn from them? And if so, what do they reveal about our beginnings?

Next, we will explore moments that defy natural laws—events that challenge the boundaries of science and invite us to reconsider the role of faith.

Just as the butterfly effect shows how small changes can alter the course of events, our beliefs about the initial conditions of life influence how we interpret patterns, draw conclusions from scientific data, and ultimately shape our worldview.

CHAPTER FOURTEEN

The Initial Condition of Life

SEAN CARROLL, A DISTINGUISHED theoretical physicist and philosopher, describes the nature of science and mathematics. He notes, "Science is a technique, not a set of conclusions."[1]. . . "Math is all about proving things, but the things that math proves are not true facts about the actual world. They are the implications of various assumptions. "[2]

Was There a Beginning?

Scientists, theologians, and philosophers generally agree that the universe had a beginning, though most theories incorporate eternity. Origin theories boil down to two possibilities: life was designed by intelligence or arose through chemical interactions on its own. Variations within each abound, but patterns in nature can guide further research.

Theories involving an intelligent designer often propose an eternal Creator who initiated our time-bound universe. Non-designer theories also include concepts of eternity, though these have evolved over time. In the past, atheists, such as Aristotle, believed the earth and our universe were eternal. Edwin Hubble's 1929 discovery of an expanding universe[3] challenged this view, leading to the rise of Big Bang theory and, for many, a shift away from the idea of eternity.

Recently, the concept of eternity has resurfaced among non-design theorists through hypotheses like Stephen Hawking's "bounce" theory, which proposes a cycle of contraction and explosion before the Big Bang. However, these ideas remain speculative, offering no explanation for what triggered the explosion or caused a prior universe to contract.

Every origin theory holds its own mysteries. Questions like "Where did God come from?" or "What set the Big Bang in motion?" lead to similar answers from both naturalists and creationists—each claim their starting point, whether God or pre-Big Bang energy, is eternal. Some truths remain beyond our grasp, defying explanation or complete understanding.

Physicist Sean Carroll describes the Big Bang as "the end of our theoretical understanding."[4] Whether we call it a beginning or a theoretical endpoint, all theories mark a start to the universe as we know it. For simplicity, I'll refer to this moment as the "beginning," whether it was sparked by a cosmic explosion or an intelligent designer.

Why discuss the beginning? Because, as chaos theory shows, initial conditions shape future outcomes. Similarly, beliefs about life's origin shape our understanding of the world and the freedom we seek. Recognizing that everything is connected—like a rhythm—helps us make sense of the patterns in our lives.

Both primary origin theories—intelligent design and naturalistic processes—face unanswered questions. Highlighting these gaps isn't to discredit any theory but to acknowledge the mystery inherent in all. Let's explore these foundational theories, keeping in mind their relevance to the patterns and rhythms of life.

Please Note: A glossary at the end of this book provides a quick reference for key terms encountered throughout this text. Most of you know the definition of a scientific law vs. a hypothesis or theory, but if it has been a while, the glossary can help refresh these definitions.

CHAPTER FIFTEEN

The Big Bang Evolutionary Theory

BIG BANG THEORY BEGINS with a singularity—an infinitely dense point of energy that defies mathematical models, thought experiments, and current physics. This concentrated dime-sized ball of energy, over a trillion degrees hot, exploded. String theory attempts to address the miracle required to explain this infinite density, but string theory remains speculative and unverified.[1]

After the explosion, the universe expanded as a "mist" of energy reflecting light, spreading faster than light itself. The hypothesized phenomenon of inflation, an unstable form of energy of an unknown nature, caused uneven clumping, pulling subatomic particles together. This process transformed energy into mass, causing the formation of planets, stars, and galaxies. Eventually, inflation disappeared—or maybe continues in the unsubstantiated multiverse. Without inflation, the mist would have continued to expand, growing farther apart until the universe reached heat death without ever clumping together.

The hypothesis of inflation serves to fill the gap between the scientific disparity of the rapidly expanding gas of the early universe and the subsequent formation of celestial bodies. Fundamental laws of science, such as the law of inertia, describe an observable pattern:

objects in motion will continue in motion unless influenced by an external force. The hot gases expelled by the Big Bang would, according to this law of science, continue to expand, growing farther apart, unless acted upon by another force to alter their speed or trajectory.

The special theory of relativity (E=mc²) states that mass and energy are conserved—they can transform into each other but cannot be Created or destroyed. Transforming energy into mass typically requires nuclear reactions or comparable processes, like those carried out at CERN[2] in the Large Hadron Collider. The conversion of energy, initially distributed throughout the early universe like mist, into all celestial bodies observed today, represents an extraordinary event. Such a conversion demanded an immense force.

Another such extraordinary event is the origin of mitochondria. Scientists hypothesize that at least 1.45 billion years ago, two cells merged, forming mitochondria, which produce the energy necessary for complex life. This "breath of life" is crucial to abiogenesis (life from non-life), yet mitochondria's formation has never been observed or replicated, nor can it be explained, making it, by definition, a miraculous event.

Nobel laureate George Wald acknowledged abiogenesis as seemingly impossible, but he argued it must have occurred because life exists. In *Scientific American,* he wrote, "The spontaneous generation of a living organism is impossible. Yet here we are as a result, I believe, of spontaneous generation."[3] However, his reasoning—we are here, so evolution must be true—is circular. This "proof" of spontaneous generation is neither scientific nor evidence-based.

Biochemist Michael J. Behe, in *Darwin's Black Box,* describes life's cellular complexity as "irreducible," arguing life's building blocks are interdependent—removing or altering any part would render the system nonfunctional. He notes scientific literature offers no explanation for the development of foundational systems such as the immune system.[4]

The Creation/formation of mitochondria is a hypothesis whose spontaneous formation defies the properties and laws governing the observed natural world. C.S. Lewis described a miracle as an event that has neither occurred before nor has any scientific precedent for its occurrence. According to this definition, the spontaneous generation of life is a miracle/singularity. Even some atheistic scientists acknowledge the Creation of mitochondria as a miracle, not in a theological sense, but as a recognition of its improbability and the unknown mechanisms involved.

Some believe the similarity between mitochondria across living organisms, from amoebas to fungi to humans, is proof of a single ancestry.

> Like eukaryotes [organisms whose cells have a nucleus] themselves, mitochondria appear to have arisen only once in all of evolution. The best evidence for the single origin of mitochondria comes from a conserved set of clearly homologous and commonly inherited genes preserved in the mitochondrial DNA across all known eukaryotic groups.[5] ~ William F. Martin, Ph.D., & Marek Mentel, Ph.D.

While Martin and Mentel believe that similarities in mitochondria point to a common origin from a single cell, this does not prove a single-cell origin. Alternative conclusions exist, but all of them are beliefs about why there are similarities, not proof.

Some evolutionary variations, such as panspermia, propose an extraterrestrial origin for life, delivered to Earth by comets, asteroids, or aliens. However, this hypothesis merely shifts the improbability of abiogenesis to another location, without addressing how life arose there. Speculations of alternate universes or a cosmological multiverse further attempt to fill gaps, yet these remain unverified philosophical ideas.

Could a multiverse exist outside of our current scientific findings? Yes, of course. Anything could, even God.

In his book *Evolution from Space: A Theory of Cosmic Creationism*, Fred Hoyle asserts aliens must have planted life on Earth because of the outrageous improbability of abiogenesis.

> No matter how large the environment one considers, life cannot have had a random beginning. Troops of monkeys thundering away at random on typewriters could not produce the works of Shakespeare, for the practical reason that the whole observable universe is not large enough to contain the necessary monkey hordes, the necessary typewriters, and certainly the waste paper baskets required for the deposition of wrong attempts. The same is true for living material.[6]

> The trouble is that there are about two thousand enzymes, and the chance of obtaining them all in a random trial is only one part in $(10^{20})^{2000} = 10^{40,000}$, an outrageously small probability that could not be faced even if the whole universe consisted of organic soup.[7]

String theory and inflation allow for a hypothetical multiverse, but these speculative concepts attempt to address evolution's gaps without reducing the incalculable improbability of life forming by chance. This reliance on miracles without empirical evidence resembles the gambler's fallacy—assuming repetition will increase the odds of success despite consistent probabilities. The gambler believes if he just throws the dice enough times, he will surely beat the odds. However, the probability is the same for every throw of the dice. At least the gambler is justified in knowing he holds all the numbers needed for the desired outcome and therefore the probability of a winning hand exists. He is not expecting a cube of six sides to transform into an octahedron of eight sides by the time it hits the table. This magnificent transformation of the dice, if thrown enough times, is what evolutionary probability is suggesting.

Even after mitochondria, millions of new traits would need to emerge for life to evolve into complex forms. The probability of such emergence does not increase with more time, no matter how much time is thrown at it. Complicating the theory even further is the lack of fossil records to support gradual emergence of new genetic data. It has led some scientists to propose leaps in evolution rather than a continuous process. However, such leaps further amplify the fantastical improbability. Returning to the fundamental building blocks of life, there is no life without mitochondria. Steven Zuryn, a molecular geneticist, describes the mitochondria as "the breath of life."

> Our primordial ancestor was a simple single-celled creature, living in a long-term rut of evolutionary stagnation. Then something dramatic happened—an event that would literally breathe life into the eventual evolution of complex organisms [the emergence of mitochondria].[8] ~Steven Zuryn

CHAPTER SIXTEEN

The Intelligent Designer

THE THEORY OF INTELLIGENT design proposes that an intelligent being Created the laws of science, the universe, and life itself, breathing life into all living beings. According to this view, the singularities required for life converge in a deliberate act of Creation by an intelligent designer, who is considered the architect of science and the laws we discover. Proponents cite the observation that life descends rather than ascends, and they highlight the failure of experiments to demonstrate abiogenesis. Scientific evidence also fails to provide adequate support.

Intelligent design primarily addresses origins, but interpretations of the designer's identity vary. Some proponents attribute the intelligent designer to God, while others suggest extraterrestrial intelligence or leave the identity undefined. Believers who attribute design to God differ in specifics—some support a young Earth Created in six days, while others propose an Earth billions of years old shaped by divine intervention. Regardless, proponents assert numerous scientific discoveries support intelligent design, and furthermore, intelligent design aligns with established evidence in the field.

Some interpretations closely align with Big Bang theory, suggesting an intelligent designer initiated the singularities needed for the universe's formation. This interpretation considers this designer to be the driving force behind inflation and the complexities of life, including mitochondria and the development of new traits.

Proponents argue the formation of atoms, our ecological systems, and life require intelligence rather than random processes. The formation of DNA is viewed as deliberate rather than evolutionary happenstance.

> Then the Lord God formed a man from the dust of the ground and breathed into his nostrils the breath of life, and the man became a living being. Genesis 2:7, NIV

Intelligent design theories highlight the intricate interdependencies in nature and the artistic patterns we observe. They propose life and the natural laws governing the cosmos originated as fully operational systems. Each life form was Created with a distinct role within its ecosystem, equipped with the necessary traits to sustain the larger rhythm of life.

According to this perspective, these systems would exhibit "apparent age"—mature ecosystems, established functional patterns, visible starlight, and fully operative natural laws from the start. In other words, the chicken came first—not the egg. Trees would be placed on earth, not just seeds. Dating methods, such as carbon and radiometric dating, rely on assumptions about initial conditions that do not account for a Created age.

If an ecosystem and components of the Earth were Created as a fully functioning system, then there would already be carbon in the soil to nourish plants; crystals would already have their shape in a cave. A dating method that places a 4.4 billion-year age on a crystal assumes it began from nothing. Similarly, carbon dating hypothesizes past carbon levels based on an assumption that everything originated from expanding energy. These dating methods, built on such assumptions, are then used to "prove" the age of the objects being studied.

Regardless of the age of the earth, carbon dating, which is used on organic matter, cannot track specimens older than 50,000 years. This dating method relies on the half-life of carbon-14 (^{14}C).[1] Anything older than 50,000 years would lack sufficient carbon for reliable measurement. Projecting dates within this time frame relies on theoretical reconstruction of carbon levels for the past 50,000 years. Accuracy depends on theories about historical carbon saturation in the atmosphere, which can change. Recent articles in the academic journal of the National Academy of Sciences (PNAS)[2] and the *Smithsonian Magazine*[3] suggest ongoing carbon emissions might compromise carbon dating accuracy by 2100, illustrating the dependency of this dating method on stable carbon saturation. Essentially, carbon dating relies heavily on theories of prior unmeasured atmospheric conditions. It's a bit of circular reasoning—the theory of evolutionary development builds the carbon level estimates, which we then use to test age.

Radiometric methods, such as potassium-argon and uranium-lead dating, are similarly based on assumptions of initial conditions. Intelligent design theorists suggest these methods cannot definitively prove or disprove the origins of life, as they rely on theoretical reconstructions of past events and the hypothesis that everything evolved from chemicals.

Intelligent design often highlights recurring patterns in nature, such as similarities in mitochondrial DNA across species. Proponents liken this to brushstrokes in a painting, where repeating motifs suggest a single artist. Just as art experts identify Rembrandt by his style, intelligent design theorists attribute life's patterns to a singular Creator, akin to God's brushstroke in nature—the greatest work of art.

Canopy theory, another perspective, suggests an ice shield once enveloped Earth's atmosphere, creating a tropical environment worldwide. This canopy would have mitigated harmful solar rays and distributed heat evenly worldwide, creating tropical conditions on the entire landmass. Mist would have evenly watered the continent (before continental shifts) instead of rain. The canopy theory posits dinosaurs had an abundant food supply on this tropical earth to support their massive size. Lifespans would have been longer without the harmful rays of the sun, and reptilian dinosaurs that grew continually throughout their lives would have been enormous.

Proponents suggest the collapse of the canopy—possibly due to asteroid impacts—led to a global flood. The impact could also have triggered volcanic activity and drastic climate shifts leading to an ice age. As evidence, they point to marine fossils found atop Mount Everest and within many sediment layers, along with numerous ancient flood accounts.[4]

This catastrophic event caused mass destruction, nearly eradicating all life, and burying large quantities of organic life. Theoretically, this process could have created some of today's oil reserves.

The extraordinary phenomenon of life is a miracle. The conclusions we draw often depend on our perspective: some attribute this wonder to randomness, while others see the hand of an intelligent designer. Both views offer interpretations of the evidence—the remnants of life before us. Each view requires faith in the explanations provided.

CHAPTER SEVENTEEN

The Breath of Life

WHILE IN QUEENSLAND, AUSTRALIA, Trin and I visited Thomas, a scientist I had met on an expedition to Antarctica. He had been purchasing large tracts of land to establish a wildlife corridor for cassowaries, helping to reconnect fragmented patches of their natural habitat along the coast. Thomas and his wife also ran a small, unofficial sanctuary, rescuing animals in need and caring for them as best they could.

Evenings were spent swapping stories of past adventures—tales of snakes, floods, and even chickens in the bathtub. Their property on Mission Beach was stunning. As Thomas led us on a walk around the land, we stopped by the river and watched a baby crocodile gliding effortlessly through the water with slow, deliberate sways of its tail. Unsure of where its mother might be, we stepped back from the water and continued on, making our way toward an old building now used as a barn.

There, we met Daisy—the goat.

She was clearly in distress. One of her udders had burst, and the other was swollen, indicating severe mastitis. She needed milking, but Thomas, who had been caring for her, said her milk wouldn't flow.

Daisy had given birth to stillborn kids, and now, her health was failing. She lay down, exhausted. I spoke to her in soothing tones while gently massaging her udder to stimulate milk flow. After a long massage, milk streamed, and Daisy sighed in relief, resting her head. I paused occasionally to give her breaks, but when her groans shifted from exhaustion to discomfort, I stopped and moved in front of her. Gently lifting her head into my lap, I stroked her and whispered soft words of comfort.

Then she shuddered violently. "Hold on, Daisy," I urged, looking into her eyes. She met my gaze, her distress palpable. She groaned again, and I felt her body seize. I saw the moment life left her eyes. Her breath stopped, and her heart ceased to beat. I laid her head back down on the earth, stroked her gently and whispered an apology. As I did, I heard the call of a magpie nearby. I imagined the bird commenting on Daisy's death.

I washed the filth and gangrene from my hands at the outdoor sink, then trudged back to the house to deliver the sad news that Daisy hadn't made it. That night, Thomas buried her.

Though we can medically describe death, there is a tangible shift that occurs when life leaves the body—something in the eyes changes, a distinct separation between life and non-life. Life is precious and extraordinary—a truth acknowledged by every origin theory.

Origin Theories: Similarities and Miracles

Both major theories of origins—whether intelligent design or evolution—propose that something extraordinary happened to make life possible. They both involve catastrophic events, and each theory acknowledges singularities—miracles where scientific laws are broken. Researchers have yet to discover evidence for these miracles, and observations do not support the phenomenon of life from nonlife. New genetic data does not develop in known species. Individuals often choose the theory that aligns with their preconceived notions about our position in the universe and the existence of intelligence greater than themselves.

Picture this: two friends are sitting in a kitchen having a cup of coffee. All the windows are closed and the blinds are drawn; neither friend can see outside. They hear the front door open and three of their kids run into the kitchen soaking wet. One friend said, "See, I told you it was raining!" The other friend responded, "No, they just ran through the sprinklers." The fact that the kids are wet proves neither statement. To determine which statement is true, we must look at other facts.

Belief in God is not inconsistent with scientific laws or discoveries. Intelligent design is the conclusion numerous origin scientists have come to based on the complexity of biological life.

Science, at its core, is not about proving theories but about explaining what we observe and describing the outcomes of experiments. True scientists admit conclusions are never infallible.

Science Doesn't Prove

Observation and experimentation are the roots of science. The origin of life, which was not observed, necessitates exploration through experiments and historical science—investigations that weigh the merits of different hypotheses.

In the seventeenth century, Jan Baptist van Helmont claimed that if you placed a dirty shirt and wheat in a container and left it undisturbed for twenty-one days, mice would appear. He believed that the sweat from the shirt contained an "active principle" that transformed the wheat into living mice.[1] It was widely believed at the time that life could arise from non-living matter. However, scientists like Francesco Redi, Lazzaro Spallanzani, and Louis Pasteur conducted experiments that refuted abiogenesis, consistently demonstrating biogenesis—the principle that life comes only from other life.

Stanley Miller's 1950s experiments, which simulated Earth's

hypothesized early atmosphere, produced amino acids, the building blocks of proteins.[2] These experiments fell short of creating even the building blocks for nucleic acids or the complexity needed for life, especially mitochondria—the powerhouse for life. Scientists conducted multiple experiments attempting abiogenesis, but all failed. Even Jeffrey Bada's later experiments only yielded amino acids.[3]

Despite the best efforts of some of the most intelligent scientists in the world, all experiments that have attempted to produce abiogenesis have failed. **The pattern of life is that life begets life.** Scientists either believe that life miraculously developed against all scientific findings, or that a more intelligent being Created it.

The existence of God Himself parallels this challenge. No one has seen God, and experimentation cannot produce evidence of His existence. But "absence of evidence is not evidence of absence."[4]

Rigorous testing, repeatability, and verification under controlled conditions are necessary to elevate a hypothesis to the status of a theory. Therefore, the creation of mitochondria is a hypothesis across all origin theories because we have never observed it and there is no empirical evidence to support it. The scientific community universally acknowledges the Creation of mitochondria as a miraculous event.[5]

The Pattern of Life

It is common for people to misuse the term "proof" in scientific discourse. Scientific evidence merely reveals patterns and descriptions that point to truth. For instance, the similarity of mitochondria across the most complex life is one fact—one point. Just like the children we referenced earlier who came into the house soaking wet, we needed more data to find out why they were wet. We must look at additional discoveries of life and scientific laws that can help point us to a conclusion—in this case, ascent or descent.

This helps us discover the pattern of life and the initial conditions that drive our beliefs.

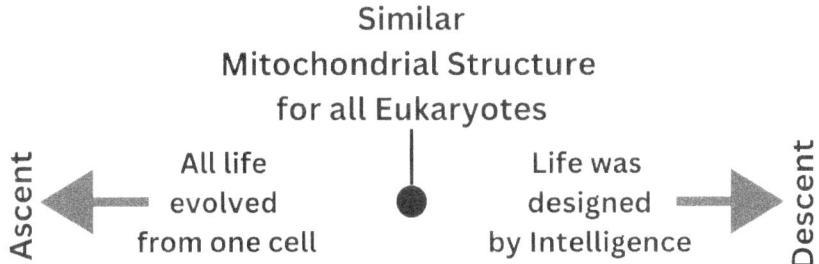

Figure 17.1 Theories of mitochondria require other observations and experimentation to strengthen the direction the pattern of life points us to.

Theories of origin serve as foundational elements in our belief systems. What we believe about our origins and ancestors influences our perception of and interaction with the world.

Let's incorporate additional scientific facts into this commonality of mitochondria to further discern the direction the arrow is pointing.

CHAPTER EIGHTEEN

The Creation of Novel Traits

MANY INFLUENTIAL FIGURES CLAIM that reliable scientific methods have robustly established the theory of evolution, which proposes that life developed from non-life following a massive explosion. In this chapter, we will delve into just one essential aspect of Big Bang origin theory, that of novel traits. The aim of this analysis is not to prove or disprove the evolutionary theory, but to contextualize this theory and scrutinize the purportedly "robust" evidence.

A novel trait refers to a genetic modification that eventually leads to new abilities and traits in a species. It is the Creation of new genetic data not inherited from an ancestor. The development of new genetic data in DNA is essential to the theory of evolutionary improvement, from a single mitochondrion cell to the complexity of life as we know it today.

Observing the evolution of gene mixing and mutations today provides insights into the complexity of life. We can simplify gene mixing by thinking of it like a lettered padlock with five unique dials labeled A, B, C, D, and E. If there are no repeated letters on a padlock, there will be 120 permutations. This single padlock has a complexity of 120 distinct patterns. Repeated letters can increase the complexity, but the original five letters still serve as the limitation. No matter how much we mix and match, Z will not be in any of the permutation options. Introducing a new element, like the letter Z, would be analogous to adding a novel trait.

Similarly, if we have five baking ingredients, we can create complexity in our bakery by mixing and matching them, but we would never have a blueberry muffin if blueberries were not one of the five initial ingredients. The initially available ingredients limit the complexity of the food we can produce. The theorized primordial soup, with its few basic building blocks, had a restricted range of complexity. It lacked the data to develop anything on its own. Novel traits are the Creation of new genetic information that was not previously available. Hence, Creation of novel traits becomes imperative at millions of junctures in the path of Darwinian evolution in order to generate the observed complexity in life forms today.

An additional illustration of this concept is the possibility of human parents with black hair having a child with blonde hair. This occurrence is probable if both parents possess a recessive blonde gene inherited from their ancestors. It is impossible for any combination of individuals to produce a child with purple hair, as purple pigment is not a trait available within human DNA for hair. Purple hair would require the Creation of new genetic information that is not present in any human.

Below is one of the most prominent examples cited as proof of the theory of ascent by a natural process. Ascent refers to the idea that life Creates additional genetic information over time. Additional examples most often used in lectures and discourse are in the Appendix. Knowledgeable professors and scientists who are proponents of spontaneous life theories frequently use these examples.

The Peppered Moth Evolution

Sewall Wright, a highly distinguished and award-winning geneticist, hailed the evolution of the peppered moth as "the clearest case in which a conspicuous evolutionary process has actually been observed."[1] The alterations in the peppered moth's appearance in Manchester, England, serve as a prominent example in numerous lectures and books detailing the evolution of life.

The peppered moth has a speckled wing resembling pepper grains. The concentration of the "pepper" ranges from light to dark colors. In 1822, observers noticed that most of the moths had a lighter color, and a minority were dark colored.

During the Industrial Revolution, the trees in Manchester became covered in soot, killing off much of the light-colored lichens where the moths often rested. By the late nineteenth century, light-colored moths became rare, and the dark color prevailed. Years later, when the environment was cleaned, the light-colored moths regained dominance.

These environmental changes appeared to influence the predominant color of moths. Although some initially speculated that birds might target less-camouflaged individuals, observations from cameras recording bird behavior seemed to challenge this theory. The primary predator of moths is the bat, which hunts at night using echolocation. Some speculated that the shift in the numerical dominance of light-colored versus dark-colored moths might be linked to the decline of lichens during the industrial stage. The exact cause is uncertain. Regardless of the reason, what we observe is plasticity. Both variants persisted before, during, and after the environmental shifts, reflecting a robust genetic pool. Plasticity entails changes in dominance within an existing genetic pool, not Creation of something new.

The dominance seen in the peppered moth cannot be used to support the hypothesis of adaptation that involves the creation of new traits. The moths did not generate new characteristics or develop new genetic data. Traits pre-existing in their DNA changed dominance when industrial soot impacted their environment.

Complexity and Pure Breeds

Some theorists argue that today's diverse species reflect an evolving world, but genetic inheritance studies suggest otherwise. Complexity often results from breaking something down into its individual parts.

An illustration of genetic loss leading to complexity, perhaps more familiar to us, is evident in the breeding of domestic animals. Breeders select dogs with specific traits to develop distinct breeds. Once established, two purebred black Labrador Retrievers will never produce pit bull offspring. Before selective breeding, ancestral dogs carried the genetic diversity found in both a black Labrador and a pit bull. By selecting for certain traits, other genetic possibilities are inevitably eliminated. For instance, breeding for short hair requires the removal of genes for long hair. This selective process reduces genetic diversity and often contributes to health issues in purebred dogs.

The complexity we see—the vast number of unique breeds—arises from isolating specific genetic traits, not from creating new ones. In fact, as each purebred lineage becomes more specialized, its ability to produce genetically diverse offspring diminishes. This is not an example of evolutionary ascent but rather of descent—the passing down and restriction of genetic information. While selective breeding allows for variation, it does not generate new genetic data. Each purebred retains only a fraction of the genetic potential of its ancestors.

The consequences of genetic narrowing are evident in breeding challenges. Over time, extreme size differences, such as those between a Chihuahua and a Great Dane, make interbreeding difficult and, in some cases, dangerous. By eliminating traits, we also impose biological limitations, sometimes even endangering the survival of offspring.

Geneticists Nicola Nadeau and Chris Jiggins, in their publication *A Golden Age for Evolutionary Genetics? Genomic Studies of Adaptation in Natural Populations*, explored recent advancements in genetics. They observed that studying recent evolutionary changes highlights the loss of traits rather than the emergence of new ones. Despite decades of research into evolutionary processes, they stated that our understanding of the genetic changes required for the development of novel traits is surprisingly limited.[2]

Complexity, in this context, is the mixing of genetic information available within the gene pool of the ancestor. The ancestral dog had complex genetics; descendant breeds are like fragments of a mosaic, each representing a small part of the original complexity. New genetic information is not being Created—rather, what we observe in our current world is a loss of genetic diversity within breeds, a gradual narrowing of the gene pool.

Genetic data is inherited information; it is not created by non-life, nor does life on Earth generate entirely new information. Even as cognitive human beings, we do not Create new information—we discover it, learn it, and rearrange it, but we do not Create new scientific information.

The pattern of information and of life is descent.

CHAPTER NINETEEN

Darwin and the Galápagos

YEARS AGO, I WATCHED a documentary on the Galápagos Islands and was captivated by the idea of an isolated haven filled with species found nowhere else on Earth. This archipelago—comprising thirteen major islands, six smaller islands, and over a hundred islets and rocks—boasts the highest rate of endemism[1] in the world. An astonishing 97% of its reptiles and land mammals, 80% of its land birds, 30% of its plant life, and 20% of its marine species exist only there. It's easy to see why Darwin was so fascinated by these remarkable creatures; they represented something extraordinary.

At the time of watching the documentary, the Galápagos felt like a distant realm, a paradise beyond my reach. But when Trin and I set foot on Baltra Island, a surge of excitement coursed through me. I nudged him eagerly and exclaimed, "We're in the Galápagos!" Our mission was crystal clear: to witness the abundance of endemic animals. Our hearts yearned to see the vibrant blue-footed boobies, the waddling red-footed boobies, the ancient marine iguanas, the playful penguins, the majestic flightless cormorants, the gentle giant tortoises, and so much more. Anticipation filled the air as we eagerly awaited the opportunity to swim alongside the legendary hammerhead sharks known to frequent these waters.

The outer islands are accessible only by boat, so we boarded a 72-foot yacht that accommodated sixteen passengers and left from the Itabaca Channel. That evening as we slept, the yacht navigated toward the remote island of Genovesa, a seven-hour journey over tumultuous seas. The crashing waves against the boat's metal hull occasionally roused us from our bunks, nearly tossing us out of bed. Once, the sound of breaking glass echoed through the vessel's halls. Yet, despite the rough waters, we found ourselves lulled into a restful slumber, swaying as if in a comforting hammock.

As morning dawned, I stumbled over to the porthole in the head and peered outside. Flocks of birds resembling miniature pterodactyls adorned a towering cliff beside the boat. These seabirds gracefully glided through the air and gathered atop the cliff on dry branches. They were the notorious frigatebirds, known for their thieving ways. They were unable or unwilling to fish on their own, shadowing other birds, stealing their catches. When our yacht set sail again, they trailed behind, riding our wake in anticipation of discarded lunch scraps. The males were a fascinating sight as they showed off their vibrant "red beards" in pursuit of a mate.

We spent that day exploring Genovesa Island, a birder's paradise that was soon to become one of our favorite islands. Stepping ashore at Darwin Bay, a multitude of red-footed boobies and Nazca boobies, unfazed by our presence, greeted us. We observed them mating, tending to their young, and strolling along the beach with their lumbering side-to-side gait. As I leaned in to capture a photo, one particularly curious booby tilted its head toward me, seemingly as intrigued by me as I was by him.

Genovesa Island is home to an array of unique species, including flightless cormorants, Galápagos doves, and lava owls. Its pristine beaches also serve as a haven for sea lions, which leisurely basked in the sun mere feet away from us. We watched in awe as a baby sea lion nursed from its mother, undisturbed by our presence.

At the cliff's summit, a sprawling expanse of palo santo vegetation resembled a bustling bird metropolis. Feeling like guests in their neighborhood, we traversed the trail, serenaded by the melodious whistles of male Nazca boobies and the raucous calls of females. Along the path, we encountered a male booby tenderly grooming its fluffy, white-feathered offspring. Another was selecting small rocks to give as gifts to his mate. We laughed when the female looked away from the gift and up to the sky. He turned, head down, to find a better rock and we chuckled about the woes of courtship.

Sea lions seem to rule these islands, lounging about with an air of ownership. On Isabela Island, Trin and I perched on a ledge in Tagus Cove to wait as the dinghy made its way to shore to take us back to our yacht. I tried to imagine the *Beagle*, Darwin's ship, anchored in this very cove on October 1, 1835—less than two hundred years ago. As we enjoyed the sunny ledge, a sea lion approached, then stopped in front of us, eyeing us as if to say, "Hey, that's my spot." Yielding to his claim, we moved down to the beach. As soon as we were out of his way, he made himself comfortable on the ledge we had just vacated.

On another island, we encountered two sea lion pups, recently born and nestled in sandy patches, the mothers tending to their young. Nearby, the discarded placentas hinted at their morning arrivals, just before our own.

As we walked farther down the beach, Trin spotted a third female. She moaned and moved about oddly. It appeared as if she was about to give birth. We watched her with swelling excitement, waiting to witness the extraordinary moment. She began to push and grunt. And then it came out! A little too easily, I thought. It was too small for a baby sea lion. It did not move, and it did not have any appendages. The adult sea lion ambled off toward the water to go for a swim and left us staring at a large turd.

The sea lion pups proved undeniably curious, boldly approaching us without fear. Tempting as it was to scoop them up for a cuddle, we kept a respectful distance, mindful of the repercussions. Human scent left on baby sea lions could prompt the mothers to reject them, and the abandoned pups would eventually starve to death. More than once, curious pups approached us to find out what we were. It was comical to watch a full-grown adult run from an adorable pup.

The terrain of the lower portion of Santa Cruz Island comprises rugged lava rock where the air is hot and dry, like that of the Arizona desert. As we ascended to a higher elevation, the air cooled, and soon after, fog closed in around us. Along with the cooler, moist air, greenery appeared in larger and denser quantities.

Upon reaching the summit, barely eight hundred meters above sea level, we found ourselves immersed in a misty realm reminiscent of an ancient world where majestic tortoises roamed freely. We meandered along lush trails, observing tortoises of impressive size leisurely grazing on grass or wallowing in the mud. These colossal creatures, with lifespans of up to 150 years, can spend up to four hours mating.

Darwin himself duly noted the distinctions between the tortoises of Santa Cruz and those of Santiago Island. He marveled at the varying sizes and shapes of these remarkable reptiles, as well as the abundance with which they populated the islands.

The wildlife on these islands is magnificent. What struck me most about them was their interaction with humans—curious and remarkably unafraid. It's almost as if they recognize the sanctuary they have there, or perhaps they've grown to understand that humans are no longer their primary threat. Even in the water, sea lions playfully circled us initiating interaction.

Spotting a pair of Galápagos sharks added to the thrill of our underwater adventure. During a snorkeling expedition near our yacht, I noticed a Galápagos shark nearby. With a quick lift of my head from the water, I exclaimed, "Shark!" Instantly, my shipmates swam over to join me, eager to observe the majestic creature as we floated in the water above it.

The islands and the waters surrounding them are teeming with life. We embarked on two dives, one off North Seymour and the other off Mosquera. Exploring the frigid water, we found ourselves surrounded by a diverse array of marine life, including white-tipped reef sharks, garden eels, moray eels, and various fish species.

The Humboldt current that flows from Antarctica up the coast of South America to the Galápagos Islands makes the water around the islands quite cold. It is also a primary driver of the diversity of life around the islands. Strong winds that drive the current displace the warm, nutrient-poor waters, causing the colder water below, which is rich in decaying organic nutrients, to rise.

After we explored the outer islands, we spent several days on Santa Cruz Island and San Cristobal, the two inhabited islands, enjoying a few excursions. Swimming beneath a vast pod of hammerhead sharks near Kicker Rock was an unforgettable experience that underscored the magnificence of the Galápagos marine ecosystem.

San Cristobal, much like the other islands we explored, teemed with marine iguanas. Their camouflage with the lava rock is so impeccable that we had to tread cautiously to avoid accidentally stepping on them. Despite their resemblance to mini-Godzillas, they posed no threat. We found a shady spot beneath a tree and watched in awe as the marine iguanas leisurely strolled along the beach, never ceasing to fascinate us. They regulate their body temperature by basking in the sun before embarking on long dives into the chilly waters to feed on underwater algae. Sea lions, seemingly just for fun, sometimes interrupt their efforts by grabbing the iguanas' tails.

The Galápagos Islands are unique. Their abundant wildlife and unique adaptations create an atmosphere of tranquility, fostering a profound connection with nature.

We concluded our exploration on San Cristobal, the very island where Darwin began his legendary five-week expedition of the Galápagos. During one of our last days there, a Galápagos finch landed gently on my hand. I cherished the intimate encounter, remaining still as the bird rested, allowing me the opportunity to closely observe this delicate creature of renowned significance. The finch, often cited as evidence of evolution, poignantly exemplifies—as we shall see in the next chapter—the descent of a species.

CHAPTER TWENTY

Finches, Rapid Evolution, and the Limits of Change

THE FINCH, SUCH A small creature, has become an icon of evolutionary study. Evolutionary biologists Peter and Rosemary Grant conducted a study of the finches on Daphne Major, an isolated island in the Galápagos. The island's small size allowed the Grants to track and catalog every finch on the island. In 1977, a severe drought impacted the island, causing a significant drop in the finch population, from 751 birds to just 90. Before the drought, the finches' beak depth ranged from 8mm to 11mm, with an average size of 9.2mm.

The drought reduced the availability of spurge seeds, the finches' primary food source, forcing them to rely on the more challenging-to-eat Caltrop seed. This seed required a stronger beak. After the drought, the Grants measured the beak depth of the remaining 90 finches and found an average increase to 9.7mm (a 15% increase). Scientists lauded this as a dramatic jump in evolution in just one generation.[1]

The increase in the average beak depth to 9.7mm resulted from the deaths of smaller-beaked birds, with 88% of the population perishing.

The new average still fell within the original range of 8mm to 11mm. This shift did not involve the development of anything new—rather, it reflected a significant loss of birds, specifically those with smaller beaks.

If a child initially has 500 Legos, with 250 being blue and the other 250 red, the average amount of red Legos is 50%. If the child loses all 250 blue Legos, leaving only red Legos, the average amount of red Legos has increased to 100%! This increase in the average does not constitute something gained. It reflects loss. For the finches, the increased average reflects the death of the smaller beaked birds.

Five years later, between 1982 and 1983, Daphne Island experienced an unusually heavy rainy season, leading to abundant spurge seeds. The larger beaks became a liability, and the average beak depth dropped by only 2.5%. The finches did not regain their original diversity in beak depth.

Survival of the fittest resulted in the death of those whose beak size was least suited to the changing food source. It did not Create a beak size on any bird outside the parameters of the available genetic data. The observed changes resulted from genetic losses, reflecting entropy and a reduction in future genetic diversity. The findings challenge the notion of Darwinian evolution as they showcase the loss of genetic information over time—a slow death.

The finches on the Galápagos Islands were influential in shaping Darwin's ideas about the evolution of species. Yet, the study by the Grants only reveals how an isolated population can experience genetic information loss and thus become distinct from their ancestors, rather than acquiring new information that enhances their genetic makeup. The study illustrates a process of entropy, diminishing future diversity within the flock. Darwin put the arrowhead on the wrong end of the line of information. He thought they had ascended—become greater than their ancestors, developing new traits. The Grants' study, though touted as evidence for evolution, only shows descent and entropy.

Many bird species have lost the ability to fly, and separated populations may undergo natural selection that eliminates traits, such as a strong beak, that are unnecessary in their specific environment. Over time, these separated populations may even lose the ability to mate with other birds descended from the same ancestor. Using the analogy of a padlock discussed earlier, losing a genetic element—such as the letter E in a sequence of A, B, C, D, and E—reduces the possible permutations from 120 to just 24, significantly decreasing complexity.

The original finch from which the Galápagos finch descended likely possessed a rich amount of DNA. As flocks separated and encountered different environments, natural selection and environmental needs led to entropy and genetic loss of traits that were a liability in their environment.

The Galápagos Islands are a unique place in this world, as if they are a step back in time. Darwin likely felt this unique atmosphere during his exploration.

Isolated from South America by 900 km (560 miles) and distanced from their ancestors by many generations, the robust animal population on the islands has changed, with less genetic diversity than their ancestors possessed. Intriguingly, they might have preserved specific genetic traits that their mainland ancestors shed. It is as if the islands have shaped their own purebred finches, eliminating liabilities—yet there is no evidence that anything new has been Created.

This study by the Grants, often cited as evidence for Darwinian evolution, instead reveals a slow degradation of genetic information, one piece at a time, a march toward extinction.

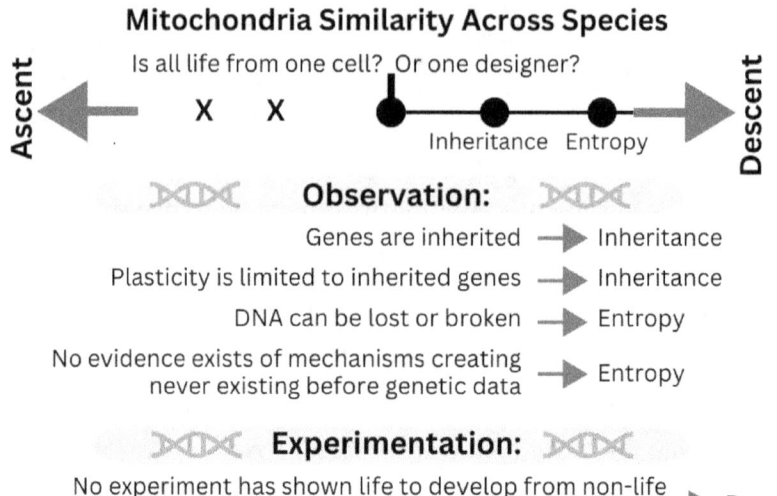

Figure 20.1 If life started from a single cell, according to the pattern of life, it would require all DNA data from every descendant life form. Alternatively, perhaps a Creator designed life, each after its kind (one ancestral canine from which all canines descended).

Death, One Detail at a Time

It has been over 150 years since Darwin first proposed the hypothesis of gradual evolution through adaptation and gene mutations. Our efforts to find evidence of changes that are not reductive have fallen short. Nor do we have the fossil records to support this hypothesis. Even ardent evolutionists like Stephen Jay Gould acknowledge the persistent absence of such evidence. He stated: "The absence of fossil evidence for intermediary stages between major transitions in organic design, indeed our inability, even in our imagination, to construct functional intermediates in many cases, has been a persistent and nagging problem for gradualist accounts of evolution."[2]

Gould, of course, is not promoting intelligent design. He is a staunch evolutionist who has hypothesized that species experience extended periods of stability with minimal change, punctuated by episodes of rapid change leading to the emergence of new species. In contrast with Darwin's concept of gradual adaptation, Gould suggests that new genetic information bursts onto the scene. This implies a series of mysterious events—instances where new life or genetics, not inherited, arise, perhaps because of necessity or stress-induced mechanisms. The emergence of new genetics is comparable to abiogenesis, with the same lack of empirical evidence.

The belief that we came from one cell and that new traits developed along the way is called Naturalism. Yet, there is nothing natural about abiogenesis or new trait Creation—at least not observed or shown in empirical evidence.

Genetic loss and plasticity account for all observed mutations and changes, with none leading to the Creation of new genes capable of inheritance by descendants.

Naturalism necessitates the emergence of millions, if not billions, of novel traits for life to evolve from amino acids and attain the complexity observed today. Yet, we cannot find evidence for even one.

Mathematically, complexity is constrained by inheritance. A string of five letters has only 120 possible permutations, and a single number can only generate a sequence with a complexity of one. Likewise, individual mitochondria cannot, on their own, account for the vast diversity of life unless new genetic information is introduced.

Life is more complex than a simple string of numbers. The usable information in DNA exemplifies irreducible complexity. Scientists have long debated which came first—DNA, RNA, or single proteins—yet none can explain where the coding information originated. All usable information that we observe today arises from intentional design or direct inheritance, just as computer programs don't come together by randomly assembling parts—they require intentional design. Life, which is far more complex than any computer program, also has a pattern that points to an intelligent origin.

> Once we see, however, that the probability of life originating at random is so utterly minuscule as to make the random concept absurd, it becomes sensible to think that the favourable properties of physics on which life depends are in every respect deliberate.[3]

> Any theory with a probability of being correct that is larger than one part in $10^{40,000}$ must be judged superior to random shuffling. The theory that life was assembled by an intelligence has, we believe, a probability vastly higher than one part in $10^{40,000}$ of being the correct explanation. . . . Indeed, such a theory is so obvious that one wonders why it is not widely accepted as being self-evident. The reasons are psychological rather than scientific.[4]

> ~Fred Hoyle, atheist and author of *Evolution from Space: A Theory of Cosmic Creationism*

Chapter Twenty-One

Objectivity and the Complexity of the Eye

While exploring Tasmania, we delved into the history and practices of the Palawa people, the traditional Aboriginal owners of Tasmania. Notably, they mastered the art of free diving for abalone along the island's coastline. According to an information board, this practice supposedly led to the evolution of larger bones around the ear in the Palawa people.

Experts recognize the enlarged bone structure around the ear in divers as "diver's" or "surfer's ear," and it can develop in individuals who spend extended periods in cold water, regardless of their ancestry.

Diver's ear is not an inheritable trait. Just as children aren't born with extra muscle mass because their parents were bodybuilders—traits like these require individual effort to develop.

I find erroneous information about evolution on tourist sign boards and in books all over because it is popular to claim findings of evolution. One erroneous claim after another leads the masses who follow to believe. "If a lie is told often enough, people will begin to believe it."[1]

The Amazing Eye

Cephalopods, such as squid and octopuses, possess a camera-like eye featuring a single lens and retina, resembling vertebrates but with the retina and optic nerve fibers in inverted positions. Jumping spiders, though not evolutionarily linked to cephalopods, also exhibit a similar camera-like eye with a single lens, focusing light onto the retina and boasting superior eyesight among arachnoids. Evolutionary theory attempts to explain this shared trait through convergent evolution.

The convergent hypothesis suggests similar traits emerge because of comparable environmental needs. This would require that the incalculable improbability of the complex eye spontaneously developing occurred more than once. It would be like two separate cultures developing an identical language and accent independently. Yet, what we observe is similar languages developing their own accent and even dialects as populations separate.

If the probability of the eye spontaneously developing is akin to the chance of an untrained monkey alone in a room with a typewriter accidentally producing the entire English dictionary error-free, then convergence would suggest that two monkeys completed this side by side, each producing an error-free dictionary. When we apply the same faith to future probability, it seems a lot less believable.

Others propose horizontal gene transfer (HGT), the sharing of DNA through means other than procreation, as a mechanism for such shared traits. HGT predominantly occurs in prokaryotes (bacteria and microbiomes). Prokaryotes are single-celled organisms that lack a defined nucleus and do not contain mitochondria. Prokaryotes use HGT to gain immunity to antibiotics. HGT is rarely, if ever, found in eukaryotes (fungi, plants, and animals). Even if a eukaryote somehow used HGT, it does not circumvent the challenge of explaining the origin of novel traits in the initiating species.

What makes the eye even more remarkable is that each species has a variation that is perfectly suited to its needs to survive in its specific environmental niche. This allows the creatures to complete their part in the ecological rhythm extremely well. Eyes, like the fundamental physical constraints of the universe, are finely tuned.

Charles Darwin once expressed skepticism about the evolution of the intricate eye through natural selection, stating, "To suppose that the eye with all its inimitable contrivances . . . could have been formed by natural selections [evolution] seems, I freely confess, absurd in the highest degree."[2] Then, Darwin explained his belief that if slight variations are discovered to exist and if they could be inherited, it was possible given enough time. Yet, 150 years later, we have failed to discover any of the slight variations in the proposed evolutionary chain, nor have we demonstrated any new variations that offer new inheritable genetic data.

Proponents of intelligent design theory attribute these shared traits to a Creator's fingerprint or gift, akin to an artist using the same pigment in multiple paintings. The Creator uniquely made each eye to suit the needs of the creature it was given to. These traits are identifiers of a singular designer who gave each creature what it needed to perform its role.

Mitochondria and complex genetic data are key to the rich and abundant life on Earth. Science cannot explain their origin beyond a hypothesis. The essence of life remains a realm beyond the complete grasp of scientific explanation, despite our accumulated knowledge.

Eyes are amazing. In them, we can see life, and we can see when the breath of life leaves them.

Eyes and Arrows

In *The Pattern Seekers*, Simon Baron-Cohen explores how the human drive to recognize and create patterns has driven progress. Throughout the book, he details how the human brain has evolved, which is why one passage in the book stood out to me in stark contradictory contrast. He cites the creation of a bow-and-arrow as proof of cognitive advancement in the human brain, relying on the calculations of Saravanane Carounanidy, M.D., to illustrate the complexity of the process. According to Carounanidy, at least nine distinct steps are required to make a simple bow-and-arrow, resulting in an astronomical number of permutations. Baron-Cohen concludes:

> "So, the likelihood of the bow-and-arrow having been made by chance is vanishingly small: just 1 in 362,880—or near-impossible."[3]

In other words, the bow-and-arrow required intelligence to be developed: it could not have been assembled by accident.

But if the relatively simple construction of an arrow is too improbable to occur by chance, how much more improbable is the development of something as intricate as the eye or the human brain? Life is vastly more complex than an arrow, yet naturalistic explanations insist that living systems—including cognition itself—emerged without guidance. This raises an unavoidable contradiction: if something as basic as a bow and arrow could not arise without intelligence, how can we accept that the eye, the brain, and life itself formed purely through blind processes?

The arrow was made by a human because it was needed—but need alone did not bring it into existence. Intelligence was required. Why should we believe that nature, before the rise of cognition, somehow created itself and its own intelligence simply out of necessity? The logic that denies randomness in simple tools while accepting it in vastly more complex biological systems is inconsistent at best. Believing that life arose from non-life requires faith.

An Objective Viewpoint

I do not know if extraterrestrial life exists, but for the sake of an objective discussion, let's entertain the notion that it exists and that a few extraterrestrials visit Earth. Imagine these intelligent beings exploring New York City and questioning you about its origin. Would they find your explanation plausible if you asserted that the entire city emerged over time by random chance with no intelligent input?

Assuming these aliens possess considerable intelligence, given their ability to travel interstellar distances, let's delve further into your explanation. You expound your theory by saying that ocean waves shaped clay into squares. Then the sun baked them into bricks, and powerful winds assembled them into massive buildings with heating and cooling systems. It was all a natural process. Would such an account convince these extraterrestrial observers?

Perhaps the alien is unconvinced, so you introduce the concept of millions of years, expecting it to believe that each building's foundation endured while subsequent tiers were added. Unfortunately, the alien, with its keen observations of how wind erodes and clay crumbles, might find this narrative implausible. Just as we observe over time that species lose genetic data and go extinct, entropy is a law of science. Life forms don't survive without the entire healthy ecological system working.

When we contemplate a skyscraper, we inquire about the architect, the designer, the glass manufacturer, the machinery provider, and even the brick maker. Yet, when we ask about the origin of life, which is exponentially more complex than a skyscraper, we are told natural elements formed it—with a few miracles along the way. The grandeur of nature, the intricacies and the interdependencies of life forms surpass anything intelligent humans have built in any great city. Miracles indeed are necessary.

CHAPTER TWENTY-TWO

The Complexity of Life

I ADMIRE THINKERS SUCH as Sean Carroll, a vocal atheist, for their skill in logical exploration. In *The Big Picture*, Carroll meticulously delves into fundamental questions about reality, the nature of life, and evolution. He does not overlook the evidence of entropy (loss) and seeks to explain the emergence of complex life amidst the challenges posed by the loss of transmitted data and the increase in disorder.

Carroll illustrates the growing complexity of our universe with an example involving a cup of coffee and creamer. When creamer is poured on top of coffee in a cup, it gradually dissipates and isn't very interesting. However, if a spoon is used to stir the coffee, we observe a more interesting environment in the cup—cream-colored tendrils gracefully descending and delicately swirling into the darker liquid. According to Carroll, this moment of beautiful complexity, just before the cup completely mixes and transforms into a homogeneous mocha-colored drink, mirrors our current stage in the universe's development. Two independent substances mix, displaying fractal beauty before entropy takes its course, homogenizing them into total entropy.[1]

This illustration echoes Lorenz's analogy of a drop of milk in coffee, a classic example in chaos theory. Lorenz explained that predicting the direction and dispersion of a drop of milk is impossible—a manifestation of chaos. Despite the initial unpredictability, the outcome consistently results in a homogeneous color and temperature—total entropy.[2] Carroll in his writing suggests we are currently in the chaos phase—the unpredictable and complex mixing.

One might argue that an intelligent being mixed the two substances, but Carroll addresses this by stating, "The spoon is an external influence, but not a guided or intelligent one."[3] All right, he has a point there; I'm not exactly at my most intelligent before I've had my morning cup of coffee either. Still, this brings us back to the "unknown force" necessary for life to exist. The mixing of the coffee is an act that involves energy, gravity, momentum, and an unnamed force, not to mention separately making the creamer and coffee in the first place, unless they are eternal.

The resulting complexity is constrained by the data available in the original substances—the coffee and creamer. Despite the satisfying amalgamation, the cup of milky coffee will never morph into a cup of hot chocolate, as the latter involves complexities beyond the original components.

Complexity is like the variations of snowflakes, each unique yet all made from the same molecular substance.

If a child is playing with a bag of Legos, she can construct a city with cars and people, but all of them will be plastic. By combining existing elements, diversity and complexity can be achieved, yet certain limitations persist. The same is true for genetic information. Creamer and coffee, providing a complex design before completely mixing, is an example of plasticity, where available substances mix. When genetic information mixes, as exemplified in dog breeding, we gain the appearance of complexity and diversity in a species, but entropy of genetic information is the result. So far, we have discovered no instances of genetic traits in a species that were not inherited. Just as the child with the Legos cannot Create metal rebar from plastic blocks, coffee and creamer will never be chocolate. We cannot make gold from iron.

If we contend that life's diversity is merely a result of complexity emerging out of primordial soup, we face a physical and logical paradox. Scientific experimentation and observation show that biological parents can increase the apparent complexity of their household by having children, each with unique personalities and DNA mix. However, every bit of genetic information in those children comes from their parents—it is inherited. Each child is a recombination of existing genetic data, but no new information outside that mix is Created. The permutations available reflect the richness of the genetic pool, not the spontaneous generation of novel traits.

If all life is merely an extension of that primordial soup, then all genetic information must have existed at the very beginning. In other words, for inheritance to function, the full sequence of DNA would have had to be present at the start—containing everything we observe today. This leads to an absurd improbability, one that defies chance. Unless, of course, a more intelligent being was there to make the coffee and milk the cow. Ultimately, this reasoning brings us back to a moment of Creation.

Carrol refers to our complex state as the mixing/messy middle—a phase between the universe's original simplicity, when energy was at rest and before its eventual return to simplicity in heat death,[4] where all matter and energy will again be at rest.

Interestingly, some theologians refer to life on Earth as "the messy middle."[5] though they start from an entirely different premise. Big Bang theorists frame existence between two states of death—an explosive origin followed by an inevitable end. Some theologians, by contrast, see existence as bookended by perfection: Creation at the start, a broken and chaotic middle, and ultimately, redemption.

Despite their opposing perspectives, both acknowledge our present reality as one of complexity and disorder.

Dinosaur Footprints in Bolivia

On our visit to Cal Orcko, just outside Sucre, Bolivia, my husband and I were able to see the largest cache of dinosaur footprints found in the world. Paleontologists have uncovered over 10,000 preserved prints. The exposed wall, currently around 9,000 feet above sea level, once rested below the ocean. Geological movements have pushed what was once a plain up on end so that it is now a wall. The cache of footprints walking and running across the wall looks like a dinosaur highway. It is a spectacular sight.

Some prints go in straight lines. Some dinosaurs walked alone, and others appeared to have been running. Another set looks like a family of two adult dinosaurs with a small set of footprints between them. In many places, the paths cross. This was a happening place.

Upon discovering this fossil highway, scientists flocked to the site to identify the species that made each print. After identification was complete, another set of scientists arrived who disputed one identification because "that dinosaur, according to our theories, didn't roam South America." They pushed the issue to change the original identification to match their theory.

According to the scientific method, it was the theory that should have changed, not the species identification. The religious belief in a theory, when the theory becomes more important than the evidence, prevents us from learning because, evidently, those dinosaurs roamed South America.

The Fossil Records

Stephen Jay Gould faced criticism from many colleagues for presenting the lack of fossil evidence to support gradual evolution theories proposed by Darwin. The backlash he encountered highlights the emotional attachment some scientists may have to their theories, akin to religious devotion, reminiscent of historical attempts to suppress alternative viewpoints, as seen with Copernicus and the Church and the culture of his time.

> "The extreme rarity of transitional forms in the fossil record persists as the trade secret of paleontology. The evolutionary trees that adorn our textbooks have data only at the tips and nodes of their branches; the rest is inference, however reasonable, not the evidence of the fossils."[6] ~Stephen Jay Gould

Darwin's original theory suggested gradual evolution over billions of years through mutations and adaptations that Created new inheritable genetic data. More and more scientists debunk this theory due to lack of evidence in the fossil record for any intermediary states. Some scientists now propose modifications, suggesting periodic large jumps in evolution, significantly increasing the mathematical improbability. Or maybe it was all one big jump, like the Creation of each living creature after its kind from an intelligent being in the beginning—followed by the complexity of descent from a rich initial genetic pool.

Despite ongoing scientific exploration, there is no evidence refuting intelligent design, which is consistent with new discoveries. The fundamental difference in choosing evolutionary theories based on chance and natural selection over intelligent design theories lies in personal preexisting beliefs about the existence of a higher being—or sometimes just faith in the scientists of our time.

The Pattern of Descent

Empirical evidence consistently demonstrates that studied species often begin with a richer genetic state than their later forms. A striking example comes from the population of crickets in Hawaii. A genetic mutation in a male cricket helped save his species from an introduced predator. The mutation, which resulted in female-like wings, rendered him unable to chirp, allowing him to avoid detection by the predator.

While this adaptation may have safeguarded the crickets on the island, it came at a cost—the loss of the genetic ability for any of his descendants to chirp.

The mutation was not the creation of new genetic information but rather a corruption of existing DNA. He inherited a trait already present in the gene pool (female wing structure) while losing the male genetic ability to chirp. This change reduced genetic diversity within the population rather than expanding it (full story in Appendix).

The Galápagos finches lost the range of beak depth available in their genetic pool when a drought killed off all the smaller beaked birds. This and other mounting evidence gives us more points to draw a line, and the arrow points to a pattern of descent rather than ascent, challenging the notion that species are evolving to higher states.

Observations and the laws of nature and science align with entropy, a process characterized by species extinction, DNA loss, and the absence of abiogenesis. The natural life and death cycle has a great deal of complexity within the confines of inherited genetic data, but we have not observed spontaneous improvements beyond the inheritability limits. Humans build gadgets and cities, but these constructions require intelligent input, resource availability, and continual energy to stave off their entropy.

Published Scientific Evidence

I have searched *PNAS* and other scientific journals and read books telling me that evolution occurred, but none of them show evidence that I can find. Michael J. Behe, in his book *Darwin's Black Box,* describes his extensive research to find anything in all the scientific libraries and research centers that can even provide a plausible theory on the emergence of the first life. He found stories on how one cell swallowed another, and instead of digesting, became a symbiotic organism, but they are just stories with no scientific support, no observation, no experimentation to recreate it, and no explanation on how those two cells did something different from all observable nature. None of them can explain the molecular steps needed. None of them provide examples of new traits developing that can be genetically inherited by the next generation. These stories defy the scientific findings of biology research.

Behe describes the notion of irreducible complexity as a system with distinct parts, all of which are required for the system to function.[7] Richard Lewontin inadvertently uses this same argument when discussing DNA. He stated:

> "No living molecule is self-reproducing. . . . Not only is DNA incapable of making copies of itself, aided or unaided, but it is incapable of 'making' anything else . . . the proteins of the cell are made by other proteins, and without the protein-forming machinery *nothing* can be made."[8]

Proteins cannot make other proteins without the information from the DNA.
DNA without protein would become extinct and new proteins cannot be made without DNA.
It is a co-dependent circle of life.

It is an irreducible complexity, meaning the entire system had to emerge complete—or else it was Created with design.[9] Our current understanding cannot prove either conclusion. They are hypotheses one believes.

Darwinian evolution is a theology that we are encouraged to believe based on scientific authority (with no scientific papers to back it up).

Dr. Boris Schmidtgall, Ph.D., a molecular biologist, has spent his career studying the structures many consider to be the most primitive structures—the complex molecules believed to be the basis of life. He has synthesized nucleic acids and proteins in the lab.

Synthesis involves assembling new combinations from existing components. In the case of proteins, this means constructing them from amino acids using genetic instructions. While scientists can chemically replicate and combine molecules, they have never been able to create anything truly self-replicating.

With his deep understanding of the immense challenges involved in merely copying nature, Dr. Schmidtgall has long pondered how simple molecules could spontaneously form life's building blocks and self-organize into cells. If synthesizing these compounds in a controlled lab environment is so difficult, how could they have assembled themselves by chance? His colleagues, despite their expertise, have yet to provide a satisfactory explanation either. He concluded his lecture titled "Evaluating Models of the Origin of Life" for the ELF Science Network with these reflections.

We have no findings that support the possibility of the origin of life through chemical evolution. On the other hand, there are a plethora of findings which clearly speak against this possibility. Based on the present state of knowledge, the origin of life is impossible without the intervention of a Creator.[10] ~ Dr. Boris Schmidtgall, Ph.D.

Life Is Miraculous

My point is not that science has proven the existence of God. Rather, believing that nature alone developed life—despite enormous gaps that theories like inflation and chance attempt to fill—requires a leap of faith. Science continually makes this belief less likely. Our growing understanding of natural processes does not diminish the need for a Designer; in fact, it reveals a profound intelligence at work.

We can use our understanding of natural patterns and laws to explain phenomena like solar eclipses or hurricanes, but explaining how something works doesn't diminish the existence of a Creator. Just as studying a watch to understand its intricate mechanics doesn't negate the need for a watchmaker, our ability to replicate or analyze nature doesn't erase the evidence of design.

Many who study molecular evidence conclude that the miracle of life points to intelligent design. However, not all origin scientists who reach this conclusion attribute the designer to God.

The patterns we see repeated in nature are those of co-dependence—entire ecological systems with species that cannot survive without the entire system. We observe entropy; we observe descent. It is the pattern of life.

Grappling with a belief in the existence of God is <u>not</u> a matter of choosing between science and spirituality. Belief in intelligent design coexists harmoniously with scientific discovery—in fact, it completes the patterns we observe. Belief about our origins is a personal inquiry, one that each of us must confront individually. For those inclined toward a belief in God, the subsequent reflection involves who He is. Is the Intelligent Designer God? If so, is God, the Creator of time and the patterns of this world, involved today, or is He a distant actor?

We often lean on observation and experimentation as the foundation of our beliefs. Yet, to comprehend the intricacies of existence, we must acknowledge the realms beyond scientific understanding. What are the unknown forces? Where is the edge of the universe? What truths are all the patterns of our world pointing us to? Does God exist?

Instead of searching for the theory of everything, maybe we should search for the rhythm of everything. To uncover the Truths that set us free, it is imperative to hold hypotheses with flexibility, distinguishing among facts, theories, and assumptions. Too frequently, we adhere to the opinions of authority figures or those within our social circles, neglecting the broader perspectives that contribute to a comprehensive understanding of the rhythms governing our existence.

One can't prove that God doesn't exist.[11] ~ Steven Hawking

Science is never finished because the human mind only uses a small portion of its capacity, and man's exploration of his world is also limited. If we look at this tree outside whose roots search beneath the pavement for water, or a flower which sends its sweet smell to the pollinating bees, or even our own selves and the inner forces that drive us to act, we can see that we all dance to a mysterious tune, and the piper who plays this melody from an inscrutable distance—whatever name we give him—Creative Force, or God—escapes all book knowledge.[12] ~ Albert Einstein

French naturalist Félix-Archimède Pouchet (1800–1872) argued against spontaneous generation (abiogenesis) in the French Academy of Sciences. Louis Pasteur, a microbiologist, chemist, and pharmacist known for his foundational discoveries in vaccines, was a firm believer in abiogenesis. Pasteur became even more intent on proving that life came from non-life and conducted more rigorous experiments, trying every angle and possibility he could test. As George Wald states in his article published in *Scientific American*, "When he [Pasteur] had finished, nothing remained of the belief in spontaneous generation."[13] Yet, in the same article, Wald states that he believes life arose from non-life because he considers the idea of an intelligent designer to be irrational and mystical. In other words, he practices Naturalism, a religion characterized by unwavering belief in unexplained events—miracles he instead refers to as singularities. Ironically, his dismissal and even disdain of other perspectives mirrors the very dogmatism he criticizes other religions for.

Collectively, we can all agree that life came from somewhere. Something amazing that, against all scientific laws and experimentation, Created life. All agree that life is a phenomenon. Life truly is wonderful; even Jason Mraz agrees in his smooth vocals and acoustic guitar when he sings, "Life is wonderful."[14] Is there not wisdom in the design of it all?

To help us on our quest for truth that will enable us to enter the Rhythm of Everything, let's examine what lies beyond the exactness of science. Let's listen to the rest of the music.

CHAPTER TWENTY-THREE

Interpreting the Evidence

SCIENTIFIC FINDINGS OF THE expansion of the universe predict an end somewhere far in the future, and they predicate a beginning. Life came from somewhere, but the theory of evolutionary development without intelligence is only one conclusion. Conclusions do not equal facts.

In the 1960s, John Calhoun conducted experiments with rat utopias and mouse paradises to explore the impact of population density. These studies involved creating controlled environments for rodents—big metal enclosures, providing unlimited food and water, with little individual cubbies to call home.

Initially, the rodents reproduced rapidly, but as the population density increased, aberrant behaviors emerged. Dominant bullies led to increased isolation among females, and aggression escalated to the point of fatal encounters. Calhoun observed a subgroup he termed the "beautiful ones," rodents that never showed aggression. They hid and seemed to forget how to interact or procreate. Eventually, each utopia ended in complete annihilation.

I came across articles discussing this experiment suggesting that population density could lead to the annihilation of humanity. The narrative seemed to project an unavoidable doom once reaching the tipping point in population density, but it appeared to oversimplify the complex factors involved. These articles used the study merely as a reference to support an agenda, framing it as a version of Armageddon without delving into the behavioral insights the study offered.

Headlines:

"Half a century ago, a rat theory predicts the end of civilisation [sic]"[1] —*Sciencepost*

"How 1960s Mouse Utopias Led to Grim Predictions for Future of Humanity"[2] —*Smithsonian Magazine*

Fact: Caged rat populations isolated from their natural environment died.

What if the demise of these rodent colonies was not solely due to population density, or even linked at all? Could it be that the rodents suffered from a disruption in their natural rhythm? Rodents have an innate drive to search for food and confront natural predators. Did the absence of the natural task of hunting leave them bored and anxious, lacking the exercise that hunting and fleeing predators would provide?

Without predators to contend with, did the dominant rats have any outlet for their aggression except for each other? They had no room to establish territories. In addition to the lack of space, their communication may have been disrupted, as rats also use ultrasonic frequencies. The large metal box may have interfered with their communication, each echoed sound serving as a constant reminder of their confinement. These intelligent creatures were jailed and deprived of natural stimuli: no joy of hunting, no beauty of nature, no need to build a community safe from predators, and nowhere to escape from each other. Could this explain their isolation, their huddling in corners, or their aggression toward one another? They were isolated from their natural environment and the rhythms that produced abundant life.

The outcome in each colony revealed psychological disturbance. Numerous rats huddled in masses out in the open, behaving like vacant creatures, merely surviving, not truly living. The extinction of the population ensued.[3] Even in rodents, isolation from the natural rhythm of life led to their demise.

Consider this: how many humans, even in a spacious mansion with all physical needs met, would willingly spend their entire lives without ever going outside again? Confinement without access to nature, stimulation, or meaningful work could lead to aberrant behavior. Isolation from natural rhythms can be detrimental. Oh, and add in a couple of bullies in the room at the end of the mansion. This would be more like hell than paradise.

While the outcome of the rat study was clear—extinction in every experiment—the explanations of why they all died are theoretical. Science tells us what happened but falls short of explaining why. Calhoun himself did not propose doom for humanity.

Let's differentiate between theories and facts that resulted from Rat ~~Utopia~~ Hell.

Fact: The rat population died in every experiment.

Theory 1: Overpopulation leads to mass extinction with no return after a tipping point.

> Theory 2: Isolation from natural rhythms leads to aberrant behavior and eventual death.

The first theory sounds like fear-mongering and suggests a reliance on the government's need to control the population—there is an agenda behind it. The second is more nuanced with practical applications for each human. It also aligns with studies in other disciplines, such as the psychology of mass formations and how isolation leads to anxiety. Historical references have linked these mass formations to acts of genocide. By connecting the dots of multiple studies, we see a line forming and it appears to point toward anxiety and aberrant behavior as a result of severed connections.

In the 1950s, Curt Richter conducted another diabolical experiment on rats (these poor sweet creatures!), revealing that rats left to swim in a bucket with no escape options quickly drowned. However, rats rescued from that bucket and then later returned to it swam exponentially longer before drowning. They had learned to hope for a rescue. Perhaps Calhoun's rats lost all hope.

Hope and connection are both crucial for survival for all intelligent species, from rats to humans.

Part 5

The Limits of Science
and the
Melody of Fai†h

Science and math, while powerful tools of discovery, illuminate only a
fraction of the vast universe. They
can describe countless phenomena, yet they falter
when tasked with explaining love, beauty, or the profound experience
of joy. They cannot answer the enduring "why" or address the deepest
questions
that stir our hearts.

SAFE   RISKY

Consciousness Friendship

Spirituality Imagination

Gratitude

Formulas

Motivation Faith Grace Friend

Perception

SCIENCE   Worth

Laws   Infinity

Art

Worship Love Origins

Predictable   Intimacy Beauty

Devotion   Wisdom

Freedom

Abundance

Aspiration

Science offers clarity within safe, defined limits. Yet,
it remains too narrow and too cautious to guide us through the
unpredictable and messy terrain of life.

CHAPTER TWENTY-FOUR

Science—Only a Drop of Knowledge

MY LIFE CHANGED DRASTICALLY when I turned thirty. That was the year my only CD, by Jason Mraz, played on repeat in my car stereo.

Up to that point, I believed that logic and cold hard facts were enough for a fulfilling life. My youthful marriage was set for failure, starting with the notion that a partner ticking all the boxes on a naïve checklist guaranteed a happy union. Diligently fulfilling the role of a "good wife," I believed, would naturally lead to a thriving relationship—if I just gave it my all.

The marriage started rocky and grew worse each year until, after nine years, I found myself at the end of my wits and with no hope left. Reality hit like a wall of hard concrete. Something was missing that I did not yet understand.

The erroneous belief that any relationship would work if I just followed all the rules shattered, leaving me in pieces. When the marriage ended, I lost family, friends, and all I had worked toward. There was nothing left to lose, and that became my freedom.

Many years later, when Trin—my soulmate, dearest friend, and now husband—and I sold everything to embark on our journey through South America, I transitioned from a goal-driven, noise-filled existence to a new rhythm of life.

One poignant morning, in Uvita, Costa Rica, the loudest growling roar I'd ever heard jolted me awake. A congress of howler monkeys perched in the tree just outside our window. Their sound penetrated my core, filling me with wonder at these creatures. Soon, however, they swung off into the jungle. Their sound carried for miles. Eventually, the jungle quieted to the sound of a myriad of bugs and birds singing out as they completed their daily lives. I was no longer surrounded by the noise of "progress," but the sounds of abundant life. Music for the soul filled me. The relentless demands on my attention and the persistent calls of advertisements to consume more and more were finally silenced. The jungle's symphony, devoid of human demands, transported me to a place of peace. It needed nothing from me—it only gave, filling me with joy.

When the noise of "civilization" was silenced, replaced by the music of nature, I wrote my personal story, starting from childhood. Writing my first chapter in a Panamanian hacienda, the process was slow, extending over the next few years. During that time, I wrote, observed beauty, and contemplated life. Months later, in Sucre, Bolivia, I confronted the events that had shattered me. The anxiety that clung to a beautiful song about life from Jason Mraz was released. A couple of years later, in Western Australia, with the sound of the Kookaburra's laughter accompanying the sunset, I completed my journal. These years were a gift that gave me time to process and confront my own erroneous beliefs. Even though I have not published that story, writing it gave me time to accept forgiveness and embrace love. Until I met Trin, I don't believe I understood love and its ineffable qualities. The journey through South America allowed me to enjoy the rhythm of life that resides beyond the exacting knife of science. It helped me see what rules and our vast store of knowledge could not explain. I could finally see the pattern in it all.

The scientific method is a powerful and invaluable tool for guiding research and uncovering patterns in the natural world. However, even in its purest form, science has limitations. It can analyze and describe patterns, but it cannot fully capture the complete rhythm of life.

As explored earlier in this book, science excels at describing and measuring repeatable, observable phenomena. Yet, it falls short of addressing concepts like love, beauty, happiness, peace, worship, and belief. It neither proves nor disproves supernatural phenomena, such as abiogenesis or the existence of God, and it cannot predict the future. The infinite nature of irrational numbers remains unexplored, just as science has yet to map the edge of the universe or uncover the depths of our own oceans.

Science does not make moral or aesthetic judgments. While scientists can study neurons and chemical processes to understand how "feeling in love" affects the body, science cannot teach us how to give love, or how to receive it. If love were reducible to a formula, we would all follow the prescribed path and never be without it. Relationships are more thrilling and complex than any scripted equation. Their unpredictability makes them both exhilarating and, at times, daunting. Love, like music or beauty, stirs emotions deep within us that science and mathematics cannot express. It is the ultimate connection, the rhythm, that completes the patterns we see in nature.

James Gleick described it in his book, *Chaos*: "He [Lorenz, the father of chaos theory] saw a fine geometrical structure, order masquerading as randomness."[1] Love masquerades sometimes as chaotic, but it contains a delicate pattern of connection.

Beauty also defies full description within the bounds of science and even language. One can delve into endless library resources to learn about the Grand Canyon in Arizona, USA, but no amount of information can capture the awe of standing on its rim for the first time. It is an intense majesty that resonates beyond verbal and scientific expression.

Intriguingly, the φ ratio, often linked to beauty, is an irrational number. Its infinite, non-repeating nature reflects the boundlessness of beauty. Beauty itself is a quality that defies full scientific or verbal expression, yet deeply fulfills us through direct experience.

Science has given us only a drop of knowledge within the vast universe. So many mysteries remain, such as undiscovered creatures in the depths of our oceans and how birds communicate to create murmurations in the sky. We don't even know what resides in the Earth's core, hidden within and beneath the crust we stand on.

As a child, I remember poring over pictures and descriptions of the Earth's mantle and core. My science book presented illustrations of Earth's core as fact, giving me no reason then to question them. However, our deepest drilling, into our eighteen-mile-thick crust, is only 7.6 miles. We haven't even reached the halfway point of the Earth's crust—the top layer! When Russian scientists drilled 5.6 miles at the Kola Superdeep Borehole—our deepest drilling at the time—they surprisingly discovered that granite did not transition to basalt as previously theorized. Our theories, however well-informed, remain subject to revision. We know that magma exists beneath the surface, as witnessed by volcanic eruptions, but we do not know how much or how deep it flows.

I have no reason to think the center of Earth differs from what scientists have theorized. It's amazing how we can measure seismic waves from earthquakes to determine density and composition of Earth's mantle. We use gravitation and magnetic studies and modeling among other methods to theoretically "see" into the center of the earth. With the intense research, we might have a great idea of what lies beneath us—but our theory of Earth's core is not within the absolute bounds of certainty. We have not observed or sampled it. Our current theories of the center of our Earth match the patterns we see elsewhere, including what we find in meteorites. However, scientists acknowledge the geophysical model of Earth's core is a theory, and this understanding propels their research, with the hope of discovery urging them to drill deeper—we don't truly know what lies beneath us.

In our pursuit of knowledge, we may feel confidence in all we have learned, akin to young children thinking they grasp everything. Yet, the vastness of unexplored realms humbles us, reminding us that despite our extensive data accumulation, we have only scratched the surface of the mysteries that await.

As we ponder the limits of discovery, questions persist: Where is the edge of the universe? Is it a closed system? Does eternity exist? Where did life come from? Does God exist, and if so, what are His thoughts? Why is the world the way it is? These mysteries dwell in realms where human knowledge encounters inherent boundaries, leaving much that science cannot define.

In a lifetime, we can only savor a mere sampling of all there is to experience, holding but a single grain of sand in the infinite universe of knowledge.

Yet patterns can show us the rhythm of life if we just step out of the middle and release those hard parts of us that blind us. The new rhythm of travel, away from corporate life, gave me the peace and stillness to process all that had transpired. It was then I understood how wonderful it was to have those unmovable legalistic pieces of me shattered and swept away. I was free simply to be and quietly observe the world's patterns.

We have shown scientifically that we are not the center of the universe, but we still struggle with thinking we are the center of the story of the universe—or the apex of it. We must seek to understand what the pattern is showing us, not just what we want to see.

The predictability of scientific facts makes us feel safe, but it is a small box to reside in. Abundant life thrives on the risks that lay outside measurable boundaries. The symphony of life weaves knowledge and logic into its foundation, but it is love and beauty that bring vibrancy to the notes, transforming structure into melody and rhythm into music.

> Existing scientific concepts cover always only a limited part of reality, and the other part that has not yet been understood is infinite.[2] ~ Werner Heisenberg

Erwin Schrödinger, the Nobel Prize-winning physicist renowned for the Schrödinger equation, made significant contributions to quantum theory. He expressed his surprise at the inadequacy of the scientific understanding of the world around him, stating:

> I am very astonished that the scientific picture of the real world around me is very deficient. It gives a lot of factual information, puts all our experience in a magnificently consistent order, but it is ghastly silent about all and sundry that is really near to our heart, that really matters to us. It cannot tell us a word about red and blue, bitter and sweet, physical pain and physical delight; it knows nothing of beautiful and ugly, good or bad, God and eternity. Science sometimes pretends to answer questions in these domains, but the answers are very often so silly that we are not inclined to take them seriously.[3]

In a live interview, Mattias Desmet warned about the risks of relying solely on logic, stating:

> Rational understanding stumbles upon an absolute limit,

and it is beyond that limit that the essence of life situates. It's the mystery of life that transcends the rational understanding. And if you continue to build that wall around you of logical reasoning—because logical reasoning is really building a wall around you—you connect one logical idea to the other. And in this way, you isolate yourself from your environment. But as soon as you start to be humble enough, as soon as you start to become aware of the fact that your rational understanding is limited, it is as if, literally, all these logical building blocks slide away from each other a little bit, and as if the eternal music of life can go through the holes of the wall and can touch the strings of your body and your soul. And it is at that moment that you can start to resonate with the mystery of life around you, with the eternal spirit of life.[4]

Do we honestly desire a life within the confines of a knowledge-built box, shielded from the unknown or anything greater than ourselves? Life's essence thrives in the chaotic spaces beyond the limits of science—a realm that, like an unfinished symphony, remains incomplete without acknowledging the greater whole. When humanity accepted that Earth was not the center of the universe, the once-chaotic movement of the planets revealed a harmonious pattern. Similarly, when we recognize our purpose as part of something greater, the seeming chaos around us aligns, and we begin to hear life's true rhythm.

Science is undeniably thrilling, and the joy of discovery delivers a dopamine boost that energizes our curiosity. However, it falls short of filling us with lasting happiness or true freedom of the soul.

As the office closes for the holiday and the evening settles in, one might find himself in an empty apartment, quietly yearning for something intangible. Often, we deflect these thoughts and eliminate the silence by immersing ourselves in distractions—streaming a movie, scrolling social media, or some other constant barrage of noise. Gadgets clutter our lives, aiming to make tasks easier, but often just complicating our existence instead. We lead busy lives with long work hours, always striving for more, running the "rat race." Perhaps we are heading toward Calhoun's "Rat Utopia," creating our own walls of false security, hemming us in with empty echoes of our severed connections.

CHAPTER TWENTY-FIVE

The Initial Condition of Faith

SHARP, URGENT BARKS ECHOED through the campground as Luna, the resident stray dog, sounded the alarm. Whines of distress and occasional yelps punctuated her vocalization. I crawled out of my tent to investigate the source of her agitation. Luna stood by the road, her muscles tense, her hackles raised, and her tail tucked between her legs. Her gaze fixed on the nearby house.

Scanning my surroundings, I did not see any trespassers lurking, so I called Luna to our tent. Speaking in soothing tones, I hoped to ease her anxiety. Though she momentarily paused her barking to glance my way, her body remained tense with apprehension. I urged her again to come toward the tent, and she reluctantly obeyed, her eyes wide and constantly darting back toward the house until she reached our tent.

Standing beside me, she continued to tremble. I reached out, offering comfort with a gentle stroke on her head and a scratch around her ears. Encouraging her to lie down next to the tent, I continued to pet her, attempting to comfort her with the tone of my voice. As she settled, I felt her tremors subside, and her anxiety seemed to dissipate. Her eyes drooped and eventually closed as her breathing slowed.

After Luna fell asleep, I crawled back into the tent. Only a whisper-thin layer of nylon separated us, and Luna remained by my side the entire night.

Trin and I were in Elqui Valley, Chile, a popular place among those seeking the strong vortex purported to be here. However, Trin and I were here solely for the majestic mountains and the world-class observatory perched atop them. Since the weather was favorable, we opted to rent a tent on the perimeter of a local's yard outside of town. Our host, along with the only other guest who occupied his spare room, were both enthusiasts of the vortex, telling us of the special energy in this valley.

The following morning, our host shared with us his belief in a rock endowed with profound spiritual significance, presenting a photograph as evidence. In the picture, he pointed to a purple orb, declaring, "See? That is the power of the gods."

I've seen this purple orb before; it happens when the sun reflects off moisture on the lens or in the atmosphere. I try to avoid them in my photography.

The other guest told us of her night, which was as restless as Luna's seemed to be. She recounted her disturbing, vision-like dream, where a dark presence climbed onto her bed and then seemed to enter her body. She attributed the vulnerability to her having forgotten to wear her protective crystals to bed.

With a self-assured smile, she interpreted the experience positively, suggesting, "It encourages me that I am on the right path because darkness is attracted to light."

Whatever happened seemed to have frightened Luna as well.

I've always found the diverse beliefs and stories of individuals and cultures captivating. Listening to others, even when their beliefs differ from mine, offers a window into their perspectives, motivations, and convictions. That afternoon, we listened with curiosity to our companion's stories, though we politely declined her invitation to participate in a Reiki ceremony.

What excited me about being in this valley was the chance to explore the cosmos at the Pangue Observatory that evening. The prospect of uncovering the universe's hidden beauty sent a thrill through me. As Cristian, an observatory scientist, keyed in NGC 253, the telescope slowly pivoted, locking onto the Silver Coin Galaxy. Through the lens, a glittering masterpiece of distant stars unfolded against the vast canvas of space.

Traveling has provided many opportunities to learn about our world—its people, geology, and the intricate patterns of nature that endlessly fascinate me. These experiences have shown me that people hold a spectrum of beliefs seemingly as varied as the human population itself. These various beliefs deserve their own in-depth exploration—perhaps in another work. For our current purpose, we will focus on the shared belief in miracles and the universality of worship.

Indeed, every human holds faith. Understanding where our faith lies, and the foundations it rests upon, is the initial condition of the choices we make. The miracles we believe in shape our faith, and our faith in origins is just the beginning of the framework that forms our view of life and humanity.

I've found that faith in God—a Creator—is harmonious with scientific discoveries. The fine-tuned, intricate patterns we observe in every atom and life-force speak of purpose, and to me, they unmistakably point to intelligent design. Claiming there is no greater force behind these patterns takes as much faith as claiming there is one—perhaps even more. After all, believing that life spontaneously appeared against all odds requires embracing the greatest improbabilities as certainties. Amidst what sometimes appears to be chaos, I see an underlying rhythm that is exquisitely beautiful, with patterns that point us to a higher truth.

So we choose the basis of our faith, the origin of life.

If we say that we believe in Science—as though it were something to be believed in—we are left with countless questions that science cannot answer, or even understand. After we study and observe, we must choose whom or what to believe—they all begin with miracles.

Intelligent design isn't about filling in those gaps by assigning a name to an unknown force or an actor for the miracles. It's about observing the patterns themselves and asking what truths they point us to. At its core, science helps us study these patterns and uncover the symmetries that govern the universe. Physicists rely on symmetry to discover natural laws, and those same patterns point to intentionality and design.

Every living thing plays a distinct role in the health of its ecosystem, reflecting a recurring theme of purpose. The wolves in Yellowstone, once thought to be a menace, have helped restore a vital life-cycle. Time after time, we observe the domino effect of removing one species: eliminating a single life form inevitably disrupts the balance, revealing the finely tuned purpose of each organism within its environment. Each is intricately designed to fulfill a specific role.

Our technological advancements have allowed us to peer deeper into the cellular world, where we find structures more complex than we could have previously imagined. Some believe that given enough time, such complexity could arise naturally. Yet, the observable pattern is that time produces entropy and extinction. The extinction of genes and of a species invariably harms, rather than benefits, its ecosystem.

Despite the clear pattern of descent, some still believe the fantastical story that life spontaneously arose and evolved. They claim that a single-cell organism "figured out" how to develop highly complex structures like mitochondria. Yet, even the most cognitively advanced species, humans, have never Created life in a lab. Still, many believe that a far simpler organism accomplished this feat, perhaps assuming it happened by chance—as if a skyscraper could spontaneously assemble itself amidst a series of storms. Naturalism believes that our intricately interconnected world emerged without guidance, evolving into the abundance and beauty we see today.

Extinction leads to greater extinction—it does not bring forth new life forms. Time breaks things down; it doesn't build them up—that's actually a scientific law. We know that progress doesn't accidentally happen. It requires intelligence, energy, and effort to construct and maintain. Yet, we convince ourselves that life arose and advanced through random chance. Scientific discoveries reveal that species descend from their ancestors, with complexity set within the parameters of inherited genetics—never beyond them. This is the observable pattern of life.

While we take pride in our scientific advancements, believing they represent the pinnacle of human achievement, we've fallen for another illusion—the belief that knowledge and measurements can answer everything. We imagine that because we understand some of the universe's laws, we no longer need the one who Created those laws.

To date, no scientific discovery has diminished belief in intelligent design. On the contrary, the deeper we explore, the more extraordinary the design reveals itself to be.

Finally, evolutionary theory struggles to explain other-centered love—the kind of selfless connection that exists beyond mere survival or reproduction. The ecological interdependence we see in forests, where plants rely on animals to process nutrients and fertilize the soil, mirrors this essential pattern of connection. It points to the essence of life: the need for relationships and the purpose each connection has.

Chapter Twenty-Six

Beyond Measure—Faith, Worship, and the Human Experience

Carbon is not a man nor salt nor water nor calcium. He
is all these but he is much more, much more.[1] ~ John
Steinbeck, *The Grapes of Wrath*

Every culture harbors a belief system, even if it doesn't involve a deity. These foundational beliefs shape our perception of the world and guide our behaviors. All cultures hold beliefs, have faith that these beliefs are true, and worship whatever they believe to be their savior—God, money, technology, or love.

The world was round even when many believed it was flat and supported by a turtle. Their misconceptions did not alter the shape of the globe; it only misshaped their understanding of the universe. When humanity finally embraced the truth that we live on a little blue marble that orbits the sun, our understanding deepened. Predictable patterns of the celestial bodies once thought to be mere wanderers in the sky took shape. Chaotic movements made sense and the pattern of their orbits took shape in our understanding. Acknowledging our position in the universe opened our eyes to its rhythms and realities. Embracing Truth has the power to liberate us from the constraints imposed by misconceptions and incomplete perspectives.

The truth will set you free. John 8:32b, NIV.

The Universality of Worship

We hold beliefs and have faith in them, and we all worship. Worship is the act of dedicating oneself to an object that is revered or adored. For instance, the phrase "She worships her job" describes a woman prioritizing her career above all else in her life. Similarly, "He worships Robin" conveys deep adoration and devotion.

David Foster Wallace, a highly acclaimed author and prominent agnostic, explored this innate human act in his commencement speech at Kenyon College in 2005.

> In the day-to-day trenches of adult life, there is actually no such thing as atheism. There is no such thing as not worshiping. Everybody worships. The only choice we get is what to worship. And the compelling reason for choosing some sort of god or spiritual-type thing to worship—be it JC or Allah, be it YHWH, or the Wiccan Mother Goddess, or the Four Noble Truths, or some infrangible set of ethical principles—is that pretty much anything else you worship will eat you alive. If you worship money and things, if they are where you tap real meaning in life, then you will never have enough, never

feel you have enough. It's the truth. Worship your own body and beauty and sexual allure and you will always feel ugly. And when time and age start showing, you will die a million deaths before they actually plant you.[2]

Wallace understood that everyone holds to a set of principles and that everyone worships. What we live for becomes our god, and our hearts align with whatever we treasure.[3]

Speak to anyone long enough, and you will uncover the faith at the core of their beliefs. Observe actions to discover what is truly worshiped.

Humanity's Search for Meaning and Fulfillment

The innate desire to seek meaning is part of human nature, propelling us to find purpose in our lives and the world. Humanity, irrespective of the diverse beliefs individuals may hold, universally shares this pursuit. The founders of the Enlightenment believed there was purpose and order to the universe—it is why they researched to find the laws holding everything in order. Why search for a thing's function unless we believe it has value or vital purpose?

Although we can explain the scientific principles behind a sunset, the joy it evokes defies mathematical description. The allure of beauty and patterns eludes complete explanation, captivating humanity. There is no formula to encapsulate these profound aspects that make us feel alive. Is there a reason we see beauty? What is the purpose of beauty?

We All Believe in Miracles

Forces unknown to scientific discovery formed our finely tuned universe, and life itself is a miracle. Beliefs about the origin of life—whether intelligently Created or chance-directed—are statements of faith. Some of the most extraordinary minds in science believe(d) in God or a greater intelligent being.

In *Encounters with Einstein,* Werner Heisenberg discusses how the revolutionary scientific method introduced by Copernicus, Galileo, and Kepler was primarily theological. Heisenberg pointed out that Galileo revered nature as God's book, and Kepler expressed gratitude for seeing the beauty in God's creation.

Werner Heisenberg (1902–1976)

Theoretical physicist, Nobel Prize in Physics, and pioneer of quantum mechanics theory

Galileo argued that nature, God's second book (the first one being the Bible), is written in mathematical letters and that we have to learn this alphabet if we want to read it. Kepler is even more explicit in his work on world harmony; he says, "God created the world in accordance with his ideas of creation. These ideas are the pure archetypal forms which Plato termed *Ideas*, and they can be understood by Man as mathematical constructs. They can be understood by Man, because Man was created as the spiritual image of God. Physics is reflection on the divine Ideas of Creation, therefore physics is divine service."[4] ~Werner Heisenberg, *Encounters with Einstein*

Johannes Kepler (1571–1630)

German astronomer, mathematician, astrologer, and natural philosopher

I thank thee, Lord God, our Creator, that thou allowest me to see the beauty in thy work of creation.[5] ~Werner Heisenberg, *Across the Frontiers*

Albert Einstein, Sir Isaac Newton, Max Planck, and Erwin Schrödinger were legendary physicists whose groundbreaking discoveries shaped our understanding of the universe. Together, they stand as pillars of both classical and modern physics, driving the evolution of scientific thought—from Newtonian mechanics to quantum mechanics and relativity.

Despite their varied perspectives, they all believed in intelligent design, as reflected in their own words. While Einstein did not believe in a personal God, he was clear that he was not an atheist, often expressing his belief in a Creator who established the laws of the universe.

Each of these great minds had faith in something beyond mere chance. Below are their thoughts on the matter, shared in their own words.

Albert Einstein (1879–1955)

German-born theoretical physicist and developer of the theory of relativity

Try and penetrate with our limited means the secrets of

nature and you will find that behind all the discernible laws and connections there remains something subtle, intangible, and inexplicable. Veneration for this force beyond anything that we can comprehend is my religion.[6]
Einstein: His Life and Universe

The problem of God's existence, he [Einstein] clarified, was simply too great for human minds to contemplate. Just as a child who cannot read, knows that a book contains great knowledge and that a library is organized according to some principle he cannot work out, so an intelligent person looking at the universe must understand that there is some guiding force at work, but attempts to work out what this is or explain its characteristics will always be an oversimplification.[7]
~*Einstein: His Life and Universe*

Sir Isaac Newton (1642–1727)

English mathematician and physicist

This most elegant system of the sun, planets, and comets could not have arisen without the design and dominion of an intelligent and powerful being.[8] ~Isaac Newton, *The Principia Mathematical Principles of Natural Philosophy*

Max Planck (1859–1947)

German theoretical physicist, 1918 Nobel Prize in Physics

There can never be any real opposition between religion and science; for the one is the complement of the other. Every serious and reflective person realizes, I think, that the religious element in his nature must be recognized and cultivated if all the powers of the human soul are to act together in perfect balance and harmony. And indeed it was not by an accident that the greatest thinkers of all ages were also deeply religious souls, even though they made no public show of their religious feeling.[9] ~Max Planck, *Where is Science Going?*

Erwin Schrödinger (1887–1961)

Austrian physicist, 1933 Nobel Prize in Physics

The scientific world-view contains of itself no ethical values, no aesthetical values, and not a word about our own ultimate scope or destination, and no God, if you please. Whence came I, whither go I?

Science cannot tell us a word about why music delights us, of why and how an old song can move us to tears.

[Science can describe] . . . the moment when certain glands secrete a salty fluid that emerges from our eyes. But of the feelings of delight and sorrow and that accompany the process science is completely ignorant—and therefore reticent.

Science is reticent too when it is a question of the great Unity . . . of which we all somehow form part, to which we belong. The most popular name for it in our time is God—with a capital 'G' . . . If its world-picture does not even contain blue, yellow, bitter, sweet—beauty, delight and sorrow—, if personality is cut out of it by agreement, how should it contain the most sublime idea that presents itself to human mind?[10] ~Erwin Schrödinger, *"Nature and the Greeks,"*

Recognizing that everyone holds beliefs and engages in worship broadens our understanding and possibilities for change. It enables us to cut through the noise and identify what genuinely provides us with freedom.

Faith, beliefs, and worship fall outside the realm that science can describe. They feel risky because they exist beyond concrete proof, yet this uncertain side of life is where joy and love thrive.

To truly live, we must have the courage to embrace the unpredictable and the unmeasurable.

Science has not eliminated belief—it relies on it. Nor has it silenced humanity's innate search for meaning. It cannot replace our need for love or for God, the Creator of love and ultimate connection. Across cultures and throughout history, faith, belief, and worship remain fundamental to the human experience, expressed in countless ways.

CHAPTER TWENTY-SEVEN

Dissonant Rhythms— Disconnection and Isolation

OUR UNIVERSE OPERATES ON finely tuned laws precisely tailored to sustain life within the ecological system. Patterns in nature display deep connections and dependencies that foster abundant life. Yet, our modern culture, influenced by Naturalism and mechanistic ideologies, often disrupts these rhythms, creating disconnection and isolation.

Disrupting Natural Rhythms

When we disconnect our belief system from the truths revealed by natural rhythms, we begin to treat animals as mere machines, cultivate monoculture crops stripped of their natural defenses and nutrition, and sever vital connections within ecosystems. This weakens natural defense systems, degrades soil health, and depletes biodiversity. In the process, we harm the land, mistreat animals, and compromise the well-being of those we seek to nourish. Ultimately, we witness the unraveling of the very ecological system we depend on—and it is already in decline.

We mourn the decline of bee populations and the exhaustion of fertile land while failing to see how our interventions cause these losses. Despite the intricacies of photosynthesis that we barely understand, we act as though we can improve it, often worsening the balance of life. The result is not progress, but a culture adept at destruction, creating disconnection.

Disconnecting Ourselves from Nature

Technology is isolating us from the rhythms of the ecological fabric, replacing its inherent artistry with a pale imitation. We immerse ourselves in temporary comforts, detached from the original Creation and the profound connections it offers.

When machines replaced farm duties, humanity migrated indoors. Our insulated, hermetically sealed homes, air-conditioned cars, and climate-controlled work environments further detach us from the natural world. The Environmental Protection Agency estimates that, on average, Americans spend 90% of their time indoors,[1] trading fresh air and sunshine for the illusion of safety and comfort.

Artificial light disrupts our circadian rhythms, allowing us to work through the night. Yet, this pursuit of productivity has disrupted sleep patterns, harmed nighttime pollinators, confused migratory birds, and affected the environment in ways we don't understand. Though abundant research highlights the health benefits of sleep and spending time in nature, we continue to insulate ourselves from these rhythms, ignoring their importance.

Disconnecting From Each Other

Busyness has become a badge of honor, leaving little time for meaningful, face-to-face connections. Brief texts, lacking the depth of nonverbal communication, displace conversations. Exhausted by the week, we retreat into mindless entertainment, bombarding ourselves with more noise. Communicating behind electronic devices, we substitute genuine relationships with virtual likes.

Our homes overflow with gadgets designed to simplify life, yet they tether us to demanding jobs just to afford them. These so-called conveniences often isolate us further, insulating our souls—our very psyches—from the true sources of joy and fulfillment. The relentless pursuit of more leaves us trapped in a cycle of stress, with jobs that continually demand more from us.

In every spare moment, we have to "do" something, so we fill every second. Instead of embracing stillness, we check social media or watch reels, consuming bite-sized entertainment.

Studies, including one from *Nature Communications*,[2] show a collective shortening of attention spans, raising concerns about our ability to evaluate the information we consume. Amid the onslaught of work stress, advertisements, and distractions, silence becomes rare. Yet, it is in silence that we can restore natural rhythms and reconnect with the world.

Our modern comforts insulate us from each other and nature, sometimes even our own thoughts. Connections to nature and other human beings have proven to increase our joy and health, but we are all once removed, resting safely inside our insulated homes, communicating over devices that filter out personal touch.

We have lost touch with Creation and the One who designed it—the One who intricately formed us. Could the design of the sun at the center of our universe be a divine illustration about the place of the Son in our lives? Placing the sun at the center of the universe transformed the chaos of "wandering planets" into harmonious patterns. When we center our lives on Him, our purpose and place in the universe create patterns that make sense of our lives and the seemingly chaotic world.

Releasing Control

Embracing silence can be daunting, forcing us to confront our thoughts and feelings, but it is essential for healing. Silence and stillness give us time to meditate on what is good and true. Above all, it is crucial to reestablish connections—not just with one another and the world, but also with something greater. It offers us an opportunity to discover the rhythm of everything.

Our culture, as described by Mattias Desmet, reflects a mechanistic ideology that reduces living things to controllable machines which we can program to our will. This ideology believes that inventions can improve the human condition. It views our bodies as machines fixed with a pill or a calorie count, ignoring their natural design. Desmet warns that the quest for perfect control erodes the essence of life, leaving people sick and depressed instead of fulfilled.[3]

The belief that we have evolved into masters of our world has led us to manipulate nature rather than work in harmony with its brilliance. Our capacity for destruction spirals up with every step of progress we make. Technology, once a promise of freedom, has instead tethered us to its devices. Our castles may be grand, but inside, the culture is isolated and bound by the inevitable pull of entropy.

Mattias Desmet observed that "Mechanistic ideology always lives on credit! In the future, once perfect knowledge has been achieved and perfect technology has been mastered, it will translocate the man-machine into paradise. Yet, for now, it mainly makes people sick and depressed."[4]

Reclaiming the Rhythm of Life

Knowledge and technology themselves are not inherently problematic; rather, it is their misuse and the importance we assign to them that cause harm. This concept is like the often misquoted saying, "Money is the root of all evil." The original phrase is: "The love of money is the root of all kinds of evil."[5] The danger lies not in its existence but in our obsession with it. By prioritizing comfort and convenience over natural rhythms, we have disconnected ourselves from the beauty and truth found in nature.

We have traded the Creator's original design for a god of comfort, insulating ourselves from the patterns that point to higher truths and greater connections. Instead, we opt for the comfortable couch and mind-numbing entertainment piped to us on electronic devices, ignoring all the things that historically and scientifically bring us joy and healing. Big and little screens, like chants from minarets, remind us five times a day that we *need more*.

The constant distraction of notifications keeps our minds moving from one task to the next with never a moment to rest, never a moment to tear down that wall growing between us and joy. Thus, we surrender our freedom for the illusion of comfort and safety.

Part 6

The Healing Rhythm of Everything

**"There are more things in heaven and earth,
Horatio, than are dreamt of in your philosophy."
~ Hamlet to Horatio**[*]

Abundant life requires connection—an orchestra where every living being plays a part. Science offers only a glimpse into this grand design. I believe our Creator designed humanity for connection, and that this is reflected in the patterns of the universe.
Nature and relationships are gentle reminders of
His enduring love.

We were designed for Connection

[*] Shakespeare, *Hamlet*, 1.5.167–168.

CHAPTER TWENTY-EIGHT

The Pursuit of a Unifying Theory

STEPHEN HAWKING ONCE STATED, "Just as there is no flat map that is a good representation of the earth's entire surface, there is no single theory that is a good representation of observations in all situations."[1]

Physicists continue their quest to find this ultimate framework, known as the *Theory of Everything*. Hawking described it as a "unified theory that will include quantum mechanics, gravity, and all the other interactions of physics."[2]

Our universe operates under the influence of fundamental forces: gravity, electromagnetism, the weak force (responsible for particle decay), and the strong force (binding particles together to form matter). The theory of general relativity describes the force of gravity. Quantum mechanics describes how the other three forces function.

General relativity and quantum mechanics are the two pillars of modern physics. While each is highly accurate within its domain, they remain fundamentally incompatible at the quantum scale. Bridging this gap is one of physics's greatest challenges—one that seeks a unified theory capable of revealing how these forces truly work together.

Hawking expressed, "If we achieve this [finding the theory of everything], we shall really understand the universe and our position in it."[3]

Perhaps the patterns in nature, the fine-tuning of our universe, the beauty surrounding us, and our capacity for joy already show us our position in the universe. Could the co-dependent ecosystem show us we need a connection to the force, the Creator, that holds it all together? Maybe the symphony of nature—the cycle of death and renewal—holds a deeper meaning, one we'll explore further. Understanding this rhythm might reveal our connection to the Creator who sustains it all.

Science demands the simplest answer. But maybe it is too simplistic to think there is a greater force holding everything together—a greater and more intelligent force that designed our universe.

We accept the theories of gravity and quantum mechanics, acknowledging gaps in our understanding. Yet, we reject the idea of God because we don't understand how an all-powerful intelligence could allow the strong and weak forces of pain and joy. There are things beyond our understanding of both science and faith.

Seeking a theory of everything must include modifications to the theories that do not work together, like relativity and quantum mechanics, or a discovery of the link that bridges the gap. Scientists acknowledge we do not know how it all works together, yet some reject logical theories that could move us closer to the ultimate theory.

Is the inconceivably rare chance of spontaneous life more believable by granting it more time? Observation and science underscore the entropy inherent in all things, describing how they degrade and die with passing time—not improve. They rust and erode, and species lose genetic information with each struggle, leading to a gradual death.

"Wait," you might say. "Some people who believe in God actually believe that a man rose from the dead 2,000 years ago. That's simply not possible!" Yet isn't the development of mitochondria more miraculous than a man rising from the dead? At least the dead man had all the cells, DNA, and mitochondria necessary for life—he just needed the metaphorical breath. If God Created the universe, life, and all the laws that they follow, then raising Jesus from the dead would be easy peasy. No one who believes He rose from the dead thinks it happened naturally. They know a miracle occurred—the laws of nature were superseded for a major event in the story of humanity. We all believe in the miracle of life.

Yet, humanity asks, if there is a God, why are things such? The scientific method readily admits unknowns, acknowledging that many things elude our understanding through scientific means. Then, hypocritically, the believer in Naturalism turns around and ridicules faith in a deity because its believers do not claim to comprehend the mind of their God.

So why do we reject the concept of an intelligent designer? Is it because we haven't observed such a being? Perhaps we consider belief in God irrational, requiring tangible proof before we can accept the idea. Could our disbelief stem from resentment toward life's hardships, leading us to question why a benevolent God would allow such events? Alternatively, is our rejection based on a desire to avoid accountability to a higher power, preferring instead to be the masters of our own destinies? There are undoubtedly many reasons for skepticism. Let's examine one: the unseen.

Believing in the Unseen—Quantum Field Theory

The initial stage of the scientific method involves observation, a crucial step in constructing a scientific body of knowledge. Through observation, we engage with the world, attuning ourselves to its sights and listening to its sounds and vibrations. Yet, what if the subject is beyond our visible or audible spectrum? The question arises: "If we can't see or hear it, why should we believe it?"

According to quantum mechanics, our perception does not accurately represent the underlying reality when we are not actively observing it. We perceive particles, but they are, in reality, fields. These particles are visible packets of energy within a broader field. Our ingrained tendency to prioritize what is observable hinders a deeper understanding of reality, say quantum physicists.[4]

Quantum field theory suggests that particles exhibit specific behaviors because they are integral components of larger, imperceptible fields. Physicists believe that there are four unseen fields, each associated with a force that governs interactions within the universe.

> For centuries, scientists have sought to describe the forces that dictate interactions on the largest and smallest scales, from planets to particles. They understand that there are four fundamental forces—gravity, electromagnetism, and the strong and weak nuclear forces—that are responsible for shaping the universe we inhabit.[5] ~ NASA

Gravity, one of the most familiar forces, serves as an example. While there is debate about whether an apple fell on Newton's head, the undeniable force of gravity is undisputed. No one has observed gravity, yet its effects are clear when we witness objects falling or feel its impact on our movements.

Physicists acknowledge the existence of unseen forces despite never observing them. Scientific discovery, often fueled by observation, doesn't always entail witnessing the force or object itself, but understanding its impact or outcomes.

Max Planck emphasized during a lecture in Florence that matter exists only through an unseen force.

> As a physicist who has devoted his whole life to rational science, to the study of matter, I think I can safely claim to be above any suspicion of irrational exuberance. Having said that, I would like to observe that my research on the atom has shown me that there is no such thing as matter in itself. What we perceive as matter is merely the manifestation of a force that causes the subatomic particles to oscillate and holds them together in the tiniest solar system of the universe. . . . We must assume that this force that is active within the atom comes from a conscious and intelligent mind. That mind is the ultimate source of matter.[6]

While our reliance on visible phenomena is common, the scientific method teaches us to look beyond what we can physically see. Physicists analyze the effects of unseen forces, and chaos theorists reveal patterns in seemingly unpredictable phenomena.

We can believe in unseen forces because a physicist told us they exist and showed us the results. We believe in wind and gravity because we can feel them and see their effects. Then we see the fine-tuned universe and the incredibly designed ecosystem and reject the idea of intelligent design, not because science has disproved God, but because we have chosen to reject Him personally.

René Frédéric Thom, founder of catastrophe theory, emphasizes that true comprehension of certain aspects of reality requires empathic resonance rather than strict rationality. "That portion of reality, which can be well described by laws which permit calculations, is extremely limited."[7]

Thom also stated,

> I am of the school of those who maintain that, even in the sciences, introspection and thought experiments play indispensable roles. All major theoretical advances, in my opinion, have arisen from the capacity of their inventors to 'get into the skin of things,' to be able to empathize with all entities of the external world. It is this kind of identification that transforms an objective phenomenon into a concrete thought experiment.[8]

Many have asked, "How can you believe in a God you cannot see?" I believe because I see the effects of His unseen force and a design that is intelligent.

Chapter Twenty-Nine

Recognizing Design in the Universe

TRIN AND I DUCKED slightly to enter a corridor leading into an ancient burial mound, feeling a childlike sense of adventure as we crept toward the inner chamber, wondering what we would find there. The entrance to the Newgrange Passage Tomb in the Brú na Bóinne Valley, Ireland, comprises three large stone slabs. Two stand on end, holding one more multi-ton slab overhead. As we progressed, cool air emanating from the massive stones embraced us. Behind me, a couple, overwhelmed by the narrow confines, turned around and exited.

As we progressed into the massive mound, the stone walls grew farther apart, and the ceiling rose in a dome fifteen feet high. The sounds from the outside world grew silent, deadened by the 200,000 tons of rocks and earth enveloping us.

Then came the moment I love. Our guide flipped a hidden switch, and we plunged into total darkness as if we were in a deep cave or mine. Our eyes could no longer detect any shapes. In the next moments, our other senses took over, and we could feel the closeness of the stones. Echoes were as dead as the bones of our ancient ancestors found here.

A collective intake of breath broke the silence, our small gasps mingling as a sliver of light transformed the room. Initially, a glow illuminated the passage that led us here. Slowly, the passage took shape as the tiny beam of light widened. Soon, muted light filled the space as the beam swept through the inner chamber.

Our ancestors conceived this mound and passage to align with the winter solstice sunrise. Just once a year, sunlight travels down the corridor to bathe this room in light, a testament to their ingenuity and understanding of celestial events.

This megalithic tomb, estimated to be built in 3,200 BCE, predates Stonehenge and the Pyramids of Egypt. Standing in a chamber that has withstood 5,000 years of rain, storms, and the coming and going of the inhabitants around it felt like stepping back in time.

We visited Brú na Bóinne in October, far removed from the winter solstice, so the light was simulated. Yet, even the simulation illustrated the precise design used to build this archeological marvel. While the purpose behind this structure is unknown and we have little knowledge about the builders, the unmistakable human design dismisses any notion of natural formation.

When we encounter something clearly crafted by human hands, we rationally attribute intelligent design to it, even if we don't know the specific creators or why they built it. We understand that the wind did not spontaneously construct that building; an architect designed it, and a contractor brought it into existence. Similarly, when we unearth an ancient archaeological structure, we recognize our predecessors were the architects.

Intelligent intent is the hallmark of everything that humans have developed. An architect studies design, patterns, and materials to create attractive buildings. Artists arrange colors on a canvas to communicate; they have a purpose behind their creations. Chefs use their understanding of food and spice to create a palate-watering dinner. Is not nature a greater architectural design and artistry far above anything humans have crafted?

Considering this, how is it deemed irrational to see something greater than human hands have ever fashioned and believe in an even greater intelligent designer? Is not life more miraculous than anything our hands have made? Does not the pattern of life revealed by science point to the higher truth of intelligent design? If it is true in the simplest forms, should it not hold true on a grander scale?

Applying the logic that acknowledges intelligence in building a quantum computer, one could extend the same reasoning to the designer behind photosynthesis which is quantum computing at room temperature.

Could an intelligent designer be the force that quantum physicists are still in search of—a force that holds everything together?

Before quantum physics was even a thought, ancient writers acknowledged the force that holds all things together. "He [God] is before all things, and in him, all things hold together." ~ Colossians 1:17, NIV

Why do we break this line of reasoning when confronted with something beyond our own capabilities? Is it because we believe nothing is greater than ourselves?

I've traversed the Taj Mahal, witnessed the Golden Gate Bridge, explored Big Ben and the Houses of Parliament, beheld the Christ the Redeemer statue in Rio, and marveled at the Sydney Opera House. I've climbed the Great Wall of China and reached the deserted city of Machu Picchu. I drove around Mount Rushmore and walked through the Forbidden City. None of these wonders occurred by accident, yet their artistry is akin to a child's work when compared to the stars in the universe and the intricacy of an orchid. No human creation, regardless of its grandeur, comes close to the pristine life beneath the glaciers of South Georgia, the towers of Torres Del Paine, the dunes of Lençóis Maranhenses with their cerulean pools, the colors of Yellowstone, or the depth of the Grand Canyon. These greatly surpass any man-made marvel. No human-made creation, regardless of its size, has inspired the same awe as the sight of orcas frolicking in the Antarctic waters. Their magnificent elegance moved me to tears, overwhelmed as I was by the breathtaking natural spectacle before me. As I watched these wonders, the song of Creation resonated within me, proclaiming the glory of God.

Yet, we entertain the belief that this universe came into existence by random chance because we do not recognize its artist; we are disconnected.

Logic Alone Is Not Enough

But logic is not enough, and we certainly can't rely on emotion. Both those opting for faith in life evolving by chance and those choosing faith in intelligent design construct their own lines of reasoning to reach their conclusions. Ultimately, the decision between an intelligent design theory and the miracle of life spontaneously emerging is a matter of faith. It frequently hinges on whom or what we choose to worship. Ultimately, it is the entity we decide to surrender to and trust. Do we confine ourselves to the box of formulas and place our sole trust in human knowledge, or do we place our trust in something more significant? Granted, for most, it's a nuanced blend rather than a stark dichotomy, but at its core, faith is a choice.

CHAPTER THIRTY

The Design of Communication

OUR WORLD THRIVES ON connections, continuously unveiling the stunning intricacies of communication that go beyond words. This repetition showcases the significance of these interactions.

Whale songs can reach across entire oceans, forests hum with fungal networks linking trees, and countless creatures communicate through frequencies beyond human hearing. Though we may never decipher the messages in the songs of birds, their melodies still bring beauty. The natural world is alive with songs and messages we may never fully understand. It teems with these exchanges, pulsating with life's unseen dialogues.

Language represents just a fraction of the extensive network through which we share and understand information. Humans can express themselves beyond structured sentences. Artists share thoughts, ideas, and emotions with their creations, and music speaks to our souls.

Symmetry communicates essential information to physicists, aiding them in uncovering truths about the universe. Thus, is it irrational to perceive these intricate patterns as a message from an intelligent designer showcasing his majesty? Is this part of the healing power of the forest, reminding us of our lost connection with its designer?

"The heavens declare the glory of God; the skies proclaim the work of his hands. Day after day they pour forth speech; night after night they reveal knowledge."[1] ~ King David, approximately 1015 BCE

Complex designs and patterns draw us in with their beauty, inviting us to search and understand. Even in the mountains, there is majesty, and the rocks themselves cry out in glorious design. Patterns communicate across the universe, radiating infinite glory. Could not these patterns repeated in our ecological system be communicating the character of their Creator?

Scientists question how the various senses of each biological creature function. They study to find the purpose of every organ and feature to understand the world, seeking purpose in every observed trait.

Understanding purpose and connection—finding out how things work together—is the rhythm of discovery; there would be no point to research otherwise. This desire to seek a purpose is part of our design. This is a gentle calling from the Creator to seek Truth. Are we listening to more than just rhetoric?

Something Incredible Happened

Something incredible occurred at the beginning of our known universe—on this, we can all concur. Our ecological system operates as a rhythm where each life cannot survive without the coexisting cycles surrounding it. To believe these intricacies happened by chance, gradually or in leaps, requires significant faith in random chance and in nature's inherent intelligence to Create itself—an intelligence of inexplicable origin.[2] It's not just the probability of the development of individual creatures that is incalculable; the entire ecological system would require corresponding changes to support one another. The Daintree rainforest relies utterly on all its life forms to thrive, with many parts unable to survive independently. If a cassowary randomly developed, it would perish unless all the fruit it depends on was available for consumption simultaneously.

Of course, evolutionary ascent doesn't suggest that the cassowary just popped into existence but instead developed over time. This means that everything would have had to develop together with changes that support each other every step of the way. One would think that species seeking to survive would develop independence, not intricate dependence.

Given that the fruit also relies on the cassowary, changes would have to be exceptionally specific with their timing. Each step of evolution would require a high level of global coordination to sustain the entire cycle. It would need to be very exact.

When we see changes in nature brought on by mutations, genetic loss, or the death of a species, we also observe a negative domino impact on the surrounding life; never the other way around. The trajectory we observe is one of decline and, ultimately, extinction.

Plasticity within species enables a range of adaptations, helping them survive environmental changes. Yet, it always remains confined within the bounds of inherited genetics—genetic information so rich that it creates resilience far beyond what a simpler design could sustain.

Intelligent design necessitates belief in an unseen force. Physicists believe unseen forces hold our universe together and allow it to resonate. We deem belief in these unseen forces logical because we witness their effects. We feel the wind and witness its caress on the grasses of the fields and its mighty force in the gale of a hurricane, but no one can see wind. But if, upon observing the intricate patterns and finely tuned ecosystems, one believes the unseen force holding the universe together is an intelligent deity, our modern religion of Naturalism dismisses it as a fairy tale, even though the effects of intelligent design permeate every atom of the universe.

The conclusion of intelligent design is not only rational; it is also logical.

Logic, of course, still resides close to the limited mathematical box. Freedom is the risky step to the right of the exacting box of science. Perhaps the greatest freedom is connection to our Creator, who made a rhythm that provides abundant life.

Finding Patterns in Chaos

At times, this world appears chaotic. We don't understand why some things happen, nor can we predict the future. But if, like Lorenz, the father of chaos theory, we take a step back to see the bigger picture, we might discover a pattern that synchronizes with the repeating patterns of connection amid a theme of death and life. These patterns work together so intricately that it is as if they are playing a symphony for us. Even before we grasp the entire song, we can sway with the rhythm.

The exquisite world is an invitation that draws us in, enticing us to search for meaning. It is a gift of joy, a message of love from a Creator to His Creation. The song it sings is the story of everything.

Have we overlooked the theory holding it all together because it defies formulation? Perhaps music is not a creation of the human brain but a gift for communicating that cannot be formulated. It's not merely a divine afterthought or an exaptation; it was designed with purpose. Some have proclaimed, "God needs to tell me He exists before I believe it." But then they ignore the communication displayed in the fabric of our world.

These rhythms offer freedom. Just as adhering to traffic rules ensures a smoother and safer journey, following the patterns designed for our well-being provides what our soul and entire being need.

God is the intelligent being who Created the laws of our universe and the knowledge that science is discovering. He is wisdom and the giver of wisdom. His thoughts surpass our understanding, yet He gives us the patterns in a stunning array for us to explore and enjoy.

We cannot see the force of gravity, yet it anchors us to our planet, pulls dying leaves from trees to be recycled into nutrients for new life, and draws droplets from clouds to nourish gardens.

> God did not leave you without something to see of Him. He did good. He gave you rain from heaven and much food. He made you happy. ~Paul & Barnabas to the people of Lystra, Acts 14:17, NLV

With each discovery, the picture widens, but the narrative remains constant, only deepening. A wonderful Creator set it in motion. All Creation can enjoy the rhythms, even if that very creation does not acknowledge its Creator. How much greater would that song be in our lives if we acknowledged and were connected to our Creator?

CHAPTER THIRTY-ONE

Extravagant Splendor—Life After the Rat Race

IN EVERY DIRECTION WE turn, majesty unfolds. It envelops us in nature, stretches beyond our telescopes, and lives beneath our microscopes. Beauty graces the minuscule crawling creatures on Earth and is magnificently displayed in the grandeur of the largest life forms. Why does such extravagant splendor abound?

Felicity Aston, the first person to ski solo across Antarctica using only muscle power, spent fifty-nine days traversing the desolate frozen desert—yet its beauty still left her in awe. She observed: "Nature is shrewd at creating perfection that is thoroughly beyond our ability to capture or replicate."[1]

In 2016, my husband and I initiated a significant life change. We tendered our corporate resignations with a four-month notice and affixed price tags to everything in our home. Strangers browsed through our belongings, leaving with vehicles full of treasures. All our possessions were for sale. The house emptied, and soon we were eating dinner perched on a yoga ball and footstool. Finally, we sold our home, packed our bags, and boarded a one-way flight to Costa Rica. The next few years saw us exploring, primarily overland, to the southern tip of South America, culminating in an expedition to Antarctica by boat. We had no deadlines.

Armed with a single backpack each and a mere week of accommodation in San Jose, our plan, when we started, condensed into one sentence: "Explore South America and, hopefully, find an empty bunk on a ship departing for Antarctica." We lived as we explored, deciding from each place where to go next and how long to stay after we arrived there. Our only internet connection was in the evenings at our lodgings (if they had it) and sometimes in the Plaza de Armas of each town. We spent most of our days disconnected from the World Wide Web. We were wherever we were and nowhere else.

Transitioning from our corporate goal-driven roles was an enormous change. In our previous professional lives, each moment demanded productivity. My action-item list perpetually expanded, even as I raced to check off tasks daily. The relentless onslaught of noise, comprising emails, news, the internet, work, and movies, left little room for respite. Even routine chores, like cleaning the house, required the accompaniment of an audiobook to satisfy my need for constant productivity. As a Type A individual, characterized by traits such as competitiveness, time urgency, and a propensity for workaholism, I found that the swift and competitive corporate pace fueled my ambition. Boarding that one-way flight to Central America felt like stepping into an entirely different life.

Our existence was simplified—no more conference calls, presentations, business trips, or performance reviews. Everything we needed fit in the packs on our backs; there were few "things" to worry about. No more lawn to mow, baseboards to dust, or junk drawers that needed to be organized at home. If we made a to-do list, it was a reasonable list that we could complete within the time frame allotted. We left behind the Big Hairy Audacious Goals and embraced a life of simplicity.

Some days we had long bus rides through the countryside. Large windows gave us a view of the passing world, whose striking drama often surprised us. We devoted many days to walking, exploring trails and towns, and unveiling a new rhythm. From the oxygen-starved Santa Cruz Trek in Peru to the snow-covered peaks of Cerro Castillo in Chilean Patagonia, each experience revealed a unique wonder. Lagoona 69 at 15,000 feet glowed with blue silt. Glacier-clad summits across the Andes Mountains pointed to ever-changing skies. The Atacama Desert displayed satin silk of mocha. Deep limestone caves in Chapada Diamantina filtered pools of water to perfection. The water was so clear that a beam of light from a hole in the cave far above pierced the darkness and illuminated the pool's sixty-foot depth as though it were blue glass. The wind created designs in the shifting sands of the endless dunes of Lençóis Maranhenses, on the edge of Brazil, constantly changing but always displaying patterns. Symmetry filled the landscape. These years became a journey of beauty and tranquility.

As these quiet years unfolded, peace came easily. Time was a gift that allowed me to immerse myself in social occasions and, when alone, to be creative.

Stepping back from the distractions of world news, social media, advertisements, and tasks that threatened to steal every moment, I found time for stillness. The voices of Western culture faded, allowing space to ponder unfolding rhythms through books, online courses, and, above all, observation.

During a poignant journey through the Villa de Leyva countryside in Colombia, I felt profound humility, contemplating the stunning landscapes and the life's journey that brought me there. From poverty and all the wrong turns in life, how was I here in a place beyond my childhood dreams? I wondered, "Who am I to have this opportunity, and why is the world so stunning?" It felt like I was brought to that moment, which allowed the patterns to finally come into focus. Beauty is more than just a function required for survival. That day, I saw the wonder and beauty of nature as a gift.

Resplendent sunsets don't seem to have a practical reason, yet they bring us joy. The ϕ ratio might serve as the most practical way for nature to arrange its petals for optimal sunshine, but it is also undeniably stunning. Fungi play a critical role in our environment, but beyond their functional significance, they are incredibly delicate, spectacular, and even enjoyable as a culinary delight.

The sun could fulfill its function without painting the sky with brilliant colors as it rises and sets. However, it creates a splendor that, if we momentarily look away, vanishes, and we miss its most brilliant displays. The collision of continents could have yielded a broken and repulsive landscape, yet instead, these very landscapes exhibit majestic power, as if the rocks themselves are crying out, telling a story of might and grace, beauty from brokenness.

Some colorful patterns in nature and bird songs serve practical purposes, such as attracting a mate. However, the question remains: why do creatures seek beauty? Some, like the colorful cleaner shrimp, cannot even see their own colors—so why are they so striking? A scent may attract a pollinator, but does the intricate diversity of each flower serve a crucial purpose for survival?

I suppose we cannot truly know what draws other creatures, as we cannot understand their emotions—only observe what they are drawn to. While their attraction may stem solely from the need for survival, one thing is certain: humanity is undeniably drawn to beauty.

Though science explains wavelengths and particles for colors, it cannot unravel the profound joy they bring. Why is there even joy? Some argue that there is no purpose or reason, attributing everything to randomness. If that's the case, why do we persistently seek to comprehend the world? Why is it considered valid to research the reasons behind mundane behaviors, such as chimpanzees throwing feces (yes, this was a funded project), and yet dismiss questioning the reason for beauty?

If there were only a sunset, if it never cast its glow on a majestic mountain top covered in snow or never reflected off a placid lake, if it were just colors in the sky, we might attribute it to some random phenomenon. But it does not confine its splendor to the sky; it filters onto all things around us. For moments each day, it transforms the way we see our world. Beauty—not just in the sky but radiating onto everything around us, filling our world—communicates and points to a higher truth, a purpose beyond randomness. Light transforms the way we see our world. The Son, the light of the world, transforms life.

Repeating patterns fill our eyes and hearts. Their symmetry permeates every viewpoint and study.

Patterns display the brushstrokes of the Creator, such as the graceful movement of birds in flight. The murmuration of flocks, as they coalesce, disperse, and harmoniously change direction, mirrors patterns found in large migrating herds and schools of fish, creating a mesmerizing display.

Philip Ball, scientist and author of *Patterns in Nature: Why the Natural World Looks the Way it Does*, marvels at the wonder of patterns. He reflects, "When we make our own patterns, it is through careful planning and construction, with each individual element cut to shape and laid in place, or woven one at a time into the fabric. . . . How does the intricate tapestry of nature contrive to organize itself, producing a pattern without any blueprint or foresight?"[2]

Despite his keen observation of patterns, Ball asserts in the same paragraph that he rejects the notion of intelligent design. He believes these patterns arise solely from the fundamental forces of nature, with no designer. Nevertheless, he remains awed by the patterns and feels they are a deep mystery.

Our culture resembles a mechanic who, after grasping the intricate workings of an engine, concludes that no designer is necessary because the engine functions perfectly on its own. Yet, the principles of nature and chemistry create patterns and systems far more complex and harmonious than any finely tuned engine—a creation for which we unequivocally acknowledge a builder. Does it not then seem reasonable to ask about the author of these natural laws and the Creator behind atoms, life, and the universe itself?

In discussing his motivation for writing *Patterns in Nature: Why the Natural World Looks the Way it Does* with Smithsonian.com, Ball explained, "You start to see patterns all around you. I hope that people will find this happening to them and that they'll appreciate how much structure surrounding us is patterned. There's just splendor and joy in that."[3]

Perhaps the symmetry of recurring beauty is guiding us toward a higher Truth. As psychologist and leadership expert Richard Farson and author Ralph Keyes, known for his work on creativity and communication, assert: "The best ideas aren't hidden in shadowy recesses. They're right in front of us, hidden in plain sight."[4]

CHAPTER THIRTY-TWO

The Rhythm of Everything

COULD THE SPLENDOR OF the natural world be a pattern that points us to something greater—the Rhythm of Everything? The ultimate question of existence surely extends beyond the whimsical 42 from *The Hitchhiker's Guide to the Galaxy*.[1] We know it is more than that.

Quantum physicists continue searching for a *theory of everything*—an equation that unites all forces and explains how the unseen binds our world together. But what if the answer is more than a formula? What if, in our myopic specialization, we forgot how it all fits together? Patterns and rhythms in nature form a grand tapestry. If we look at only one thread, its path may seem chaotic but makes sense when seen from a broader perspective.

Beauty in the Grand World of Communication

In *An Immense World*, award-winning science writer Ed Yong explores the hidden sensory landscapes of animals—realms beyond our perception. Through specialized devices, scientists have detected echolocation, electrolocation, and frequencies far beyond the reach of human senses. It's as if an entire universe exists alongside ours, unnoticed. Yong describes the experience of clipping a microphone onto a leaf to capture the world of insects as "setting foot upon an alien planet."[2]

He expands on this idea:

> "It tells us that all is not as it seems and that everything we experience is but a filtered version of everything that we could experience. It reminds us that there is light in the darkness, noise in silence, and richness in nothingness. It hints at flickers of the unfamiliar in the familiar of the extraordinary in the everyday of magnificence in mundanity."[3] ~Ed Yong

Yong's book reveals the unseen auditory landscapes of creatures and their visual experiences, surpassing our own. Engaging with his work is like discovering hidden dimensions woven into the fabric of life.

If we could hear every frequency, see the full spectrum of light, or feel electromagnetic currents like some animals, the sheer amount of stimuli would overwhelm us. Our world is noisy, yet our design allows us to find stillness and peace. Nature strikes a perfect balance between silence and sound, offering moments of tranquility while leaving us with the joy of vast discovery.

If you think nature's sounds couldn't possibly overwhelm you, try holding a conversation as a flock of sulfur-crested cockatoos screeches overhead in Australia. Or attempt to sleep as howler monkeys hold congress in the trees outside your cabin window in Costa Rica. Yet, even in these interruptions, there is beauty—a reminder of the wonder woven into creation.

We hear the cackle of the laughing kookaburra, marking the rise and fall of the sun. Music in many forms—more than mere sound—communicates emotion, stirring peace, joy, and empathy. We instinctively perceive what we need to thrive, while the rest is a gift, given for the joy of discovery. What else might we learn from these patterns?

The Dance of Life

The Daintree Rainforest thrives in seamless interdependence—trees feed cassowaries, which spread their seeds, ensuring new life. Decaying vegetation enriches the soil, turning death into renewal. Sunflowers and fungi grow with remarkable efficiency, following their inherited design. The sun fuels all life, springs nourish the land, and even winds carry Sahara dust to replenish the Amazon, linking distant ecosystems in a hidden harmony.

Is this wonderful dance of life truly the product of random chance? Could basic organisms "figure out" how to Create entirely new features that all work together in a globally interdependent system? The world is an astonishingly precise fit of elements.

The symphony of symmetry and intricate design beckons us—drawing us gently closer, stirring a longing to comprehend the profound mysteries at play. Could the language of symmetry woven throughout the universe be the fingerprint of an intelligent designer? A Creator capable of designing a world so fine-tuned could easily embed a narrative within the patterns.

There is a story written into the rhythms of every natural cycle, inviting us to see and understand. Beyond mere function, He adorned the universe with beauty—not just to satisfy our senses, but to awaken a yearning within us.

Beauty is a soft summons, calling us to seek and to know Him.

The Call of Beauty

Toward the end of Steve Jobs's life, virtuoso cellist Yo-Yo Ma visited him. A favorite of Jobs since their meeting in 1981, Ma played Bach on his 1733 Stradivarius cello. Moved to tears, Jobs remarked, "You playing is the best argument I've ever heard for the existence of God because I don't really believe a human alone can do this."[4]

Jobs was expressing not a belief in God but beauty's profound impact—something beyond mere human effort. Perhaps part of Jobs's success stemmed from his obsessive pursuit of elegance. He understood the power of art and seamless detail, integrating them into every iteration of the iPhone and Macintosh, even if he struggled with the nuances of relationships.

One of the most compelling rhythms in nature is the call of beauty. It is a language that resonates not only with our logical minds but also with our souls. The radiance of the natural world beckons us—a quiet invitation from the Creator to seek Him.

Nourishment for Abundant Life

The pattern of the forest teaches us that each organism is fed by something outside itself. A brook cannot fill itself—it depends on rain. Seeds fall to the ground and would die if not for the nourishment of rain, sun, and soil. It does not earn nutrients, it simply absorbs them and grows.

Modern culture, believing itself free from religion, has crafted its own rulebook—one that often preaches self-love as the foundation for loving others. Yet, nature tells a different story. A seed does not feed itself. We cannot fill ourselves.

When we place ourselves first, at the center, disconnection follows. But when we align with the Creator's rhythms—loving God first, then others—we step into a greater harmony. It is like the ancients who put Earth at the center of the universe and saw chaos in the heavens. When the sun was placed at the center, the patterns made sense. The Creator set His Son at the center of His story—communicated to us in the patterns of nature and in his written Word.

This connection—surrender of ourselves from the center—doesn't diminish self-worth; it affirms it. Recognizing that we were intentionally Created reassures us that our existence is not a mistake—our life has a purpose. It is a connection to the One who breathed life into us, the One who Created rain, who can fill us with joy—giving us an abundance of love to share.

Science confirms what nature illustrates: we need connection to thrive. Over and over, ecosystems reveal a central truth—death is nourishment for new life. Could it be that the death of Jesus and the miracle of His resurrection are the higher truth that all these patterns point to? His sacrifice of death brings new life. It is our path to reconnection—offering a way back to the Creator, the source of all things.

If God fine-tuned even the atom, could He not weave His truth into nature's patterns so that no one is without excuse? The world illustrates a simple yet profound reality: connection brings life, and isolation brings death.

Life Freely Given

We each choose where to place our faith, but not all beliefs reconnect us to the true source of life—the only source that can fill us. This fulfillment is not something we can earn; it must be received.

God's love cannot be earned, only accepted. Yet, acceptance requires us to dismantle the walls we build—walls of doubt, fear, and the illusion of self-sufficiency. Past wounds and the fear of surrender often keep us trapped behind them. But beyond our limited experience lies a vast and abundant universe—if we have the courage to step into it.

Receiving healing, love, and forgiveness requires humility, yet the freedom it brings is transformative. Love—unconditional love from the source of living water—is the rhythm of everything.

Ancient Voices on the Divine Patterns in Nature

Across centuries, ancient writings have recognized the natural world as a reflection of something greater. From the heavens above to the earth below, these passages describe nature as a testimony to divine design, revealing order, purpose, and wonder. Below are scriptures that illustrate how creation itself speaks of the unseen, pointing to the hand of its Maker.

Between the 7th and 4th centuries BCE

"But ask the animals, and they will teach you, or the birds in the sky, and they will tell you; or speak to the earth, and it will teach you or let the fish in the sea inform you. Which of all these does not know that the hand of the LORD has done this? In his hand is the life of every creature and the breath of all mankind." ~ Job 12: 7–10, NIV

"And these are but the outer fringe of his works; how faint the whisper we hear of him! Who then can understand the thunder of his power?" ~ Job 26:14

615–605 BCE

"For as the waters fill the sea, the earth will be filled with an awareness of the glory of the Lord." ~ Habakkuk 2:14

1015–686 BCE

"The heavens declare the glory of God; the skies proclaim the work of his hands." ~ Proverbs 19:1

CE 56

"For ever since the world was created, people have seen the earth and sky. Through everything God made, they can clearly see his invisible qualities—his eternal power and divine nature. So they have no excuse for not knowing God." ~ Romans 1:20 NLT[5]

CE 62

"We are bringing you good news, telling you to turn from these worthless things to the living God, who made the heavens and the earth and the sea and everything in them. In the past, he let all nations go their own way. Yet he has not left himself without testimony: He has shown kindness by giving you rain from heaven and crops in their seasons; he provides you with plenty of food and fills your hearts with joy." ~ Barnabas and Paul in Acts 14:15b–17

The inherent laws of nature display the rhythmic dance of the ocean and orchestrate the celestial bodies in perpetual motion. These laws give us seasons and a circadian rhythm, fostering healing as we rest during the night. Our universe is fine-tuned, and at the atomic level, particles cohere with an unseen force.

Almost any field of study can detail how it impacts every other living thing around it. Would not the Creator of such an intricately intertwined universe be far more intelligent than our imagination can fathom?

If an intelligent designer exists, His laws unmistakably echo abundance, healing, and purpose within the natural realm that surrounds us. The awe-inspiring magnificence and majesty of our world serve as a profound reflection of its designer. Would not His laws for our lives similarly embody abundance, healing, and purpose?

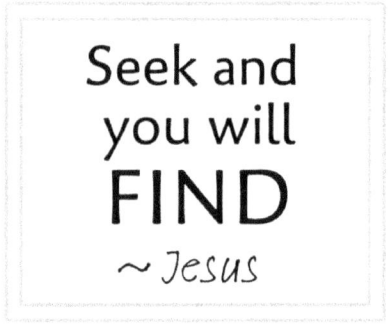

Seek and
you will
FIND
~ Jesus

Matthew 7:7[6]

Our ecological system tells the story of death that brings new life. Within the fine-tuned patterns of nature the Creator placed His story. He made it beautiful, gently wooing us to seek Him. Then He physically came to this Earth to provide the ultimate death that could give us new life—a life of forgiveness and grace, a chance to start again with a connection to Himself.

He left us more than just the story written in nature. He gave us His Word. The Bible is the greatest literary book of all time, but it is much more than literature. Many things are said of the words written on its pages, but have you read it yourself? Are you curious?

Are you tired? Worn out? Burned out on religion? Come to me. Get away with me and you'll recover your life. I'll show you how to take a real rest. Walk with me and work with me—watch how I do it. Learn the unforced rhythms of grace. I won't lay anything heavy or ill-fitting on you. Keep company with me and you'll learn to live freely and lightly. ~ Jesus, The Message[7]

Why I Believe In God

I believe in God because everywhere I look, I see His fingerprints—from the repeating patterns of the universe to the intricate details of life. I see Him in the co-dependencies of our ecosystems sustained by delicate mycorrhizal networks, in the wind and sunshine, in beauty and joy, and in the transformation of those who have surrendered to Him.

Science is extraordinary; it reveals the intricate connections of our universe. Yet, it can only describe what fits within its predefined limits. I understand that my faith lies beyond what science can explain—but science cannot negate it either. The truth has set me free, and I long to share that freedom.

Dear Reader,

I HOPE I'VE PIQUED your curiosity enough to prompt questioning of at least one of your own preconceived ideas. Frequently, it's these mistaken notions that impede our journey to freedom. If we're honest with ourselves, we all harbor a few. Personally, I'm still uncovering them within myself.

Thank you for taking this journey of listening, learning, and thinking with me. I don't judge you against standards that aren't part of your belief system or any other standards, for that matter. Such judgment is not my role. I cherish you for who you are.

For those of you struggling to survive right now, will you continue to seek? What harm could it do to reexamine your presuppositions?

My sincere wish for you is freedom.

Glossary:

Abiogenesis: A hypothetical organic phenomenon whereby living organisms emerge from non-living matter.

Adaptation: The process by which a species adjusts to or becomes better suited to its environment. Observation reveals adaptation that remains within genetic plasticity, while the theory of evolution posts that adaptation can develop new traits not inherited or obtained via Horizontal Gene Transfer (HGT).

Ascending: Moving from a lower point to a higher point, gaining information, or becoming greater. The idea of a new generation ascending biologically (with new genetic data not inherited or transferred) from a prior generation is a hypothetical idea with no supporting empirical data.

Carbon Dating: Also known as Radiocarbon dating or Carbon-14 dating, the scientific measurement of the half-life of carbon-14 within an organism. During their lifetimes, plants and animals absorb or ingest carbon from the air. Once they die, carbon absorption ceases, and the carbon acquired during their lifetime begins to break down. Scientists gauge the remaining amount of carbon in each organism based on the half-life of carbon. The age of the organism is then determined by comparing it to the estimated carbon levels in the atmosphere at the time of its existence.

Circadian rhythm: A 24-hour cyclic pattern that our bodies adhere to. The earth makes a full rotation every 23 hours and 56 minutes.

Governed by cues such as light and darkness as well as other environmental factors, it signals our brain to release hormones. These hormones play a pivotal role in regulating sleep, body temperature, and metabolism. This rhythm can vary from one person to the next.

Consilience: The convergence or agreement of multiple inductions derived from disparate sets of data, indicating a unified understanding or explanation. It denotes the act of multiple factors coming together, coinciding, or concurring to support a unified conclusion or explanation.

Creation: The act of bringing something into being that previously did not exist.

Descending: Moving from a higher place or form to a lower one. Empirical evidence has only uncovered descent in biological beings. New generations never have genetics that were not inherited from their biological parents, or in some single-celled organisms, through horizontal gene transfer. Descendants never develop new genetics on their own.

Determinism: A proposition that there is no such thing as chance. Everything has been predetermined by initial conditions. The idea of determinism (no free will) has long been a philosophical, theological, and now a scientific debate.

Entropy: The tendency toward disorder. Everything without continual input tends toward breaking down. Heat dissipates and available energy becomes less with time. (see Heat death for the state of total entropy)

Eukaryote: An organism composed of cells that have a nucleus enclosed within a membrane, along with other specialized membrane-bound organelles. This includes single-celled organisms like protists to complex multicellular organisms like plants, animals, and fungi. Conversely, **prokaryotes,** like bacteria and archaea, lack a nucleus and membrane-bound organelles, and their cellular organization is simpler.

Exaptation: A trait that has been repurposed by evolutionary development. Vestiges and exaptations differ—one refers to a trait that has lost its original function, while the other describes a trait repurposed for a new use—neither serves its initial purpose. One has theoretically been adapted for something new, while the other has diminished or disappeared.

Fractal: An infinite and repeating pattern, showcasing intricate complexity that remains self-similar across various scales. These patterns emerge through iterative processes, perpetually looping back on themselves. Nature abounds with such fractal forms, evident in the branching of trees, the meandering of rivers, the contours of coastlines, the jagged peaks of mountains, the billowing shapes of clouds, the spirals of seashells, and the swirling dynamics of hurricanes, among others.

Heat Death of the Universe: One long-term implication of the Second Law is the idea of the "heat death" of the universe. This is a hypothetical state where all energy has become evenly distributed, and no further work can be done. Entropy would be at its maximum.

Horizontal Gene Transfer: HGT is the sharing of DNA through means other than procreation. HGT is predominantly observed in prokaryotes (Bacteria and microbiomes).

Hypothesis: A conceptual idea about the natural world that is set forth as an explanation which moves science forward to prove or disprove the hypothesis. A hypothesis is not meant to be strongly held in science. It is a mode to experiment with ideas and then propose a theory, similar to a detective developing a motive from clues and then seeking proof. This approach provides a direction for further investigation.

Intelligent Design: A theory proposing that intricate biological life and the fine-tuned universe were Created by a being more intelligent than the creation itself.

Irrational Number: A real number that cannot be expressed as a ratio of two integers. It is a number whose *decimal representation* extends to the right of the decimal point *infinitely* and *without repeating.*

Laplace's demon: The hypothetical ability to predict the future by knowing every initial condition. Laplace's demon remains a theoretical idea.

Law of Entropy: A concept central to the laws of thermodynamics. Simplistically, it is described as the state of all matter and energy tending toward disorder. It is, in reality, the tendency toward thermodynamic equilibrium. A hot stone placed in a cold room will dissipate its heat into the room until both the room and the stone are at the same temperature. This is central to the theory of the ultimate heat death of our universe. Once the heat is dissipated and equal to its surroundings, there is no more energy left to make any further changes.

Law of Thermodynamics (aka Law of Conservation of Energy): A concept stating that the amount of matter and energy in the universe is constant. It can change form between energy and matter, but new energy or matter has not been observed to be Created, nor has experimentation Created new matter or energy. For example, wood can be burned, changing it from matter to energy in the form of heat and light, releasing its carbon into the atmosphere.

Mass Formation: A large group of people, sometimes an entire culture, driven by an ideology. Mass formations can be driven by a small, extreme minority, exploiting heightened anxiety and providing a target and a solution. For extended descriptions and examples, read *The Psychology of Totalitarianism* by Mattias Desmet.

Miracle: A point in time when <u>current physical laws break down</u> and known mathematical models are no longer applicable. Miracles defy the current observable and/or experimental functions or laws of our natural world. They are improbable events. This term is generally used (but not always) when the event is assumed to be directed by a deity.

Mitochondria: Tiny, double-membraned organelles found in nearly all eukaryotic cells (cells with a nucleus). Often called the **"powerhouses of the cell,"** they generate the majority of a cell's energy by converting oxygen and nutrients into adenosine triphosphate (ATP), the molecule that fuels cellular processes. In addition to energy production, mitochondria play a role in cell signaling, apoptosis (programmed cell death), and metabolic regulation. **Mitochondria DNA** is the double-stranded circular data residing within the mitochondria cell.

Mycorrhizal networks: Underground systems formed by the symbiotic relationships between certain fungi and plant roots. Essentially, these networks connect plants to one another through a web of fungal threads, known as hyphae, which extend into the soil. The fungi help plants absorb water and nutrients—like phosphorus and nitrogen—from the soil, while the plants provide the fungi with sugars and other organic compounds they produce through photosynthesis. It's a mutually beneficial partnership. These networks can link multiple plants, even across different species, where resources and information are shared. For example, a healthy tree might send extra nutrients to a struggling neighbor through the network, or plants might use it to signal chemical warnings about pests or drought. Scientists call this the "Wood Wide Web," and it's a key part of how ecosystems like forests thrive.[1]

Naturalism: The belief that everything can be understood through observation, scientific inquiry, and empirical evidence. Naturalists believe that nothing exists beyond the natural universe and reject any supernatural explanations. It closely aligns with materialism, asserting that only matter exists and that mental processes are a product of physical processes within the brain.

Novel Trait: A genetic modification that eventually leads to new abilities and traits in a species. The development of novel traits is essential to the theory of evolutionary development, from a single mitochondrion cell to the complexity of life as we know it today.

Nuclear DNA: The linear double helix of data that encodes most of the genetics in eukaryotes (humans, animals, plants, fungi).

Phenotypic plasticity: An organism's ability to change its physical traits (**phenotype**) in response to environmental conditions, without altering its genetic code (**genotype**). This adaptability allows species to survive and thrive in changing environments. For example, some plants can alter their leaf size or thickness in response to the amount of sunlight available, and some fish can change their sex based on the population's gender ratio. All observed trait changes are within the range of the organism's genetic makeup, just as a bodybuilder's muscles grow larger with more exercise. All changes remain within the available genetic data that has been inherited. No new genetics have been observed to be developed.

Science: A <u>method of investigation</u> rather than a belief system or subjective viewpoint. It encompasses both an existing body of knowledge, comprising discoveries made thus far, and the ongoing process of acquiring new knowledge through methods such as observation, experimentation, hypothesis testing, and analysis.

Scientific facts: Phenomena that have been meticulously <u>observed</u> and <u>consistently validated through experimentation</u>. They are widely accepted as reliable truths within the scientific community. For instance, the boiling point of water, which occurs at 100°C (212°F) at sea level, is a well-established scientific fact. This assertion has been substantiated through numerous experiments and observations, and deviations from this temperature occur only under specific alterations to parameters such as pressure or the presence of impurities in the water.

Scientific law: A <u>description</u> of how the natural world reacts under specific circumstances. An example is the conservation law, which states that matter and energy cannot be Created nor destroyed.

Scientism: An exaggerated reliance on scientific assertions, even when there is inadequate empirical evidence to substantiate a scientific conclusion. It is faith in human discoveries to document the truth about reality and to solve all issues in the world through a mechanistic formula.

Singularity: A point in time when <u>current physical laws break down</u> and known mathematical models are no longer applicable. Singularities defy the current observable and/or experimental functions or laws of our natural world. They are improbable events. This term is used when the event is assumed to happen by random chance.

Supernatural: Something that happens outside the bounds of the laws of nature. It is something abnormal and often considered impossible. Generally, it is believed that an unknown force caused the unnatural occurrence.

Theory: An <u>explanation</u> that incorporates scientific facts and hypotheses. A scientific theory is often based on a lifetime(s) of research, observation, and experimentation. Theories are necessary to push discovery forward, much like hypotheses. They help us create a framework that makes sense of the discovered scientific facts, but they are not absolute. Theories are educated conclusions about discovered data and should change when discovered data conflicts with them.

Appendix:

Higher Bone Density

IMAGINE POSSESSING A GENETIC mutation that results in exceptionally strong bones. While not reaching the level of Wolverine in X-Men, this mutation involves a dominant missense in LRP5 (a damaged DNA sequence), affecting the regulation of bone homeostasis. Individuals with this mutation typically experience reduced susceptibility to broken bones. While it may seem great to be so strong, medical journals classify this condition as a disease because of its detrimental effects, such as hearing loss and increased cranial pressure.

This mutation doesn't introduce new genetic information; it only decreases the abilities of available genetic data. This genetic error stems from the malfunction of LRP5 receptors; it is a syndrome affecting the body's ability to accept the protein sclerostin. Mutations and gene mixing do not produce traits that fall outside the range of combinations available within the inherited genetics, as observed by scientists. They do not Create new data. Mutations are errors that lead to a decrease in function, representing an example of the law of entropy—a gradual decline, death, one gene at a time.

HIV Immunity

A small percentage of individuals exhibit immunity or resistance to the effects of HIV, a crucial discovery aiding scientists in treating or preventing the disease. Medical research often links this immunity to a mutation in the CCR5 gene, known as CCR5-delta32, which involves the deletion of 32 base pairs from the CCR5 gene.[1] While this mutation is beneficial for HIV resistance, it represents the loss of genetic information rather than the Creation of a novel trait.

Cornish Rex Cat

The Cornish Rex cat is a highly social breed known for its soft, curly coat. It has earned a reputation for being a cat that likes to cuddle. This breed originated from a mutation in one kitten in 1950. Cats have three layers of hair, but this kitten was born with only the down coat. This cat's irresistible charm and snuggly demeanor made it a prime choice for breeding, ultimately giving rise to the Cornish Rex cat breed.

This genetic mutation, endearing to those who desire a cuddly cat, makes them more vulnerable to the cold. The need for warmth drives their penchant for cuddling. Again, this mutation does not represent an evolutionary advancement granting new abilities; rather, it increases the cats' dependence on human caregivers and reduces their ability to survive independently. They lost the genetic information to grow a warm fur coat.

Crickets in Hawaii

Another study often cited as evidence for the theory of Darwinian evolution involves a mutation observed in the Teleogryllus oceanicus crickets on the Hawaiian island of Kauai.

Male crickets on the island of Kauai once filled the air with the sounds of their wings rubbing together, creating a high-pitched chirping song to attract females. As the female approached, the song became softer and more complex, producing a sonorous rhythm. Tragically, in the 1990s, Ormia ochracea, a parasitic fly, migrated to the island, posing a significant threat to the crickets.

These flies preyed on male crickets, targeting them because their chirping made them easy to locate. Once located, the flies laid their larvae in the crickets, and upon hatching, the larvae fed on the crickets, killing their hosts.

Female crickets, however, were not targeted by this fly because they do not chirp. They have a different wing design, making them silent and therefore difficult for the ochracea fly to find. While the females remained safe, the fly maggots devoured most of the males, leading to a rapid decline in the cricket population.

Marlene Zuk studied the near extinction of the Kauai Teleogryllus cricket population on Kauai, noting their diminishing song. Over time, she discovered that the cricket population rebounded, but in silence. A genetic mutation had caused a male cricket to inherit the female wing, losing its ability to chirp. In an environment without the predatory fly, this male would most likely have died without notice, losing to the competitive chirping of the other males vying to attract a female. However, he now became the survivor because the flies could not locate him. He repopulated the island with crickets, albeit without the ability to produce sound. Natural selection, breeding with the only choice available, led to the loss of genetic data that silenced the orchestra of crickets. Mr. Silent, with a female wing, was the "fittest" to survive the invasion.

This mutation contributed to the survival of the cricket population in the face of the invasive fly species, but it represents a loss of genetic information rather than an enhancement. The crickets no longer possessed the genetic ability to chirp or produce courtship songs; that gene became extinct, reducing their future range of complexity. The soundless wing was not new genetic information, but a male with a genetically female wing. He inherited something already available in the genetic pool and he lost a valuable piece of data to the species.

Loss of genetic information is a partial death and aligns with the second law of thermodynamics, illustrating that over time, things break down or undergo entropy.

This is a step toward extinction, not new life. Even if we remove the invasive fly, the crickets will never sing again unless one singer survived and is hiding out, preserving that rhythmic piece of genetic ability.

Acknowledgments:

To my husband, Trinity, my true partner in adventure—thank you for your unwavering kindness, encouragement, and endless support as I poured countless hours into researching and writing this book.

To **Jaqueline Harrell**, whose keen editorial eye and dedication refined every page—your countless hours of editing, discussions, and thoughtful suggestions have truly elevated this book. I am beyond grateful for your expertise and generosity.

To **Amber and Jessica**, who courageously read the very first draft—thank you for your patience, insights, and early feedback that helped shape this work. And to **Jen**, who always says yes to reading my work and sharing her thoughts—I deeply appreciate you.

To **Paul Drecksler**, your encouragement and belief in this book have been a gift. Your thoughtful feedback and unwavering support have meant more than words can express.

To **Nancy Stella**, an amazing boss, mentor, and sounding board during one of my most difficult years—your wisdom and guidance helped shape not only my career but also the person I am today. I am forever grateful for your support and friendship.

A heartfelt thanks to **Tina Stephens**, whose stunning cover design captures the essence of this book so beautifully.

To **BBH and Word of Mouth Editing**, your expertise and insights polished this book into something far greater than I could have achieved alone. I'm deeply grateful for your guidance.

And finally, to **Franklin**, founder of Amplify Marketing Services and a true friend—thank you for your invaluable wisdom, industry connections, and steadfast support in navigating the world of publishing. Your generosity is a gift.

With immense gratitude,
Bonnie Truax

About the Author:

FROM A YOUNG AGE, Bonnie Truax exhibited a keen interest in the sciences. She spent her summers exploring the woods with a handheld microscope for impromptu close-up examinations. With her telescope, she studied the moon and stars, even creating a makeshift viewing box to track solar storms. These were elementary beginnings, surely, but scientific interest and study stayed with Bonnie throughout life.

Her fascination with the workings of the human mind led her to pursue a university degree in counseling, and, in an unpredictable twist, she promptly began a career in IT finance. Her proficiency in recognizing patterns and interconnected issues made her a pro at optimizing workflows. This skill became the driving force behind her professional career, eventually securing her position as a director in a Fortune 50 company.

Bonnie and her husband, Trin, retired from the corporate world to pursue life-affirming discovery of the natural world. They explore one continent at a time, opting for a one-way, long-haul flight every few years. So far, they have traveled through The Americas, Australia, Antarctica, Africa, Asia, and Europe. In each country, they enjoy delving

into forest systems, hiking geological formations, witnessing weather patterns, and immersing themselves in diverse cultures. Bonnie and Trin plan to continue their adventures, experiencing diverse patterns across the globe, for as long as they can.

Endnotes

Rhythms of the Jungle

1. Matthew Walker, *Why We Sleep: Unlocking the Power of Sleep and Dreams* (Scribner, 2017), 68.

Mysteries of the Daintree

1. A closed ecosystem is a system that sustains itself without introducing new mass.

2. Sun and wind are energy, but rain is mass, eliminating the Daintree from being a closed ecosystem.

3. Smail Mehda, Maria Ángeles Muñoz-Martin, Mabrouka Oustani, Baelhadj Hamdi-Aïssa, Elvira Perona, and Pilar Mateo, "Microenvironmental Conditions Drive the Differential Cyanobacterial Community Composition of Biocrusts from the Sahara Desert," *Microorganisms* 9, no. 3 (2021): 487. Accessed January 29, 2024. https://doi.org/10.3390/microorganisms9030487.

4. "NOAA Satellite Tracking Dust and Sand Being Blown from Sahara Desert," NOAA, June 18, 2020, https://www.nesdis.noaa.gov/news/noaa-satellite-tracking-dust-and-sand-being-blown-sahara-desert.

Life Beneath Our Feet

1. Patrick Schultheiss, Sabine S. Nooten, Runxi Wang, Mark K. Wong, François Brassard, and Benoit Guénard, "The Abundance, Biomass, and Distribution of Ants on Earth," *Proceedings of the National Academy of Sciences* 119, no. 40 (2022): e2201550119, https://doi.org/10.1073/pnas.2201550119.

2. Patrick Schultheiss, Sabine S. Nooten, Runxi Wang, Mark K. Wong, François Brassard, and Benoit Guénard, "The Abundance, Biomass, and Distribution of Ants on Earth," *Proceedings of the National Academy of Sciences* 119, no. 40 (2022): e2201550119, https://doi.org/10.1073/pnas.2201550119.

3. Proverbs 6:6–8, NIV.

4. Currently, scientists have divided slime molds into their own supergroup, but originally they categorized them within the kingdom of fungi. Slime molds also play a vital role in the death and life cycle of a forest.

5. Atsushi Tero, Seiji Takagi, Tetsu Saigusa, Kentaro Ito, Dan P. Bebber, Mark D. Fricker, Kenji Yumiki, Ryo Kobayashi, and Toshiyuki Nakagaki, "Rules for Biologically Inspired Adaptive Network Design," *Science* 327, Issue 5964 (2010): 439–442, https://doi.org/10.1126/science.1177894; Ed Yong, "Slime Mould Attacks Simulates Tokyo Rail Network," *National Geographic* (January 21, 2010), https://www.nationalgeographic.com/science/article/slime-mould-attacks-simulates-tokyo-rail-network.

6. American Association for the Advancement of Science, "Slime Design Mimics Tokyo's Rail System: Efficient Methods of a Slime Mold Could Inform Human Engineers," *ScienceDaily*, January 22, 2010, www.sciencedaily.com/releases/2010/01/100121141051.htm.

7. Jillian Kubala, "4 Immune-Boosting Benefits of Turkey Tail Mushroom," *Healthline*, published November 6, 2018, last updated April 6, 2023, https://www.healthline.com/nutrition/turkey-tail-mushroom.

8. Jennifer O. Han, Nicholas L. Naeger, Brandon K. Hopkins, David Sumerlin, Paul E. Stamets, Lori M. Carris, and Walter S. Sheppard, "Directed Evolution of Metarhizium Fungus Improves its Biocontrol Efficacy against Varroa Mites in Honey Bee Colonies," *Scientific Reports* 11, 10582 (2021): 1–10, https://doi.org/10.1038/s41598-021-89811-2.

9. Luiz A. Domeignoz-Horta, Melissa Shinfuku, Pilar Junier, Simon Poirier, Eric Verrecchia, David Sebag, and Kristen M. DeAngelis, "https://www.nature.com/articles/s43705-021-00071-7 ," *ISME Communications* 1, no. 1, (December 2021): 1-4, https://doi.org/10.1038/s43705-021-00071-7.

10. A. B. Frank and James M. Trappe, "On the Nutritional Dependence of Certain Trees on Root Symbiosis with Belowground Fungi (An English Translation of A.B. Frank's Classic Paper of 1885)," *Mycorrhiza* 15, no. 4 (2004): 267–75, doi: 10.1007/s00572-004-0329-y.

11. Yinon M. Bar-On, Rob Phillips, and Ron Milo, "The Biomass Distribution on Earth," *Proceedings of the National Academy of Sciences* 115, no. 25 (2018): 6506–6511, https://doi.org/10.1073/pnas.1711842115.

Restoring Natural Rhythms

1. Allan Savory, "How to Green the World's Deserts and Reverse Climate Change | Allan Savory," YouTube, March 4, 2013, 22:19, https://www.youtube.com/watch?v=vpTHi7O66pI.

2. Joel Salatin, *Folks, This Ain't Normal: A Farmer's Advice for Happier Hens, Healthier People, and a Better World* (Center Street, 2011), 21.

3. Brodie Farquhar, "Wolf Reintroduction Changes Ecosystem in Yellowstone," Yellowstone National Park, June 22, 2023, https://www.yellowstonepark.com/things-to-do/wildlife/wolf-reintroduction-changes-ecosystem/.

4. "Beaver," National Park Service, last updated May 1, 2023, https://www.nps.gov/yell/learn/nature/beaver.htm.

5. "The Story of Polyface Farms." n.d. Polyface Farm. Accessed January 10, 2025. https://polyfacefarms.com/about-us.

6. Joel Salatin, *The Marvelous Pigness of Pigs: Respecting and Caring for All God's Creation*, (FaithWords, 2017).

7. Joel Salatin, *The Marvelous Pigness of Pigs: Respecting and Caring for All God's Creation*, (FaithWords, 2017), "Freedom vs. Bondage."

8. Eatwild.com provides a valuable resource, listing farms across the USA that adhere to natural patterns, treating both land and animals with the care they are entrusted with.

9. Joel Salatin, *The Marvelous Pigness of Pigs: Respecting and Caring for All God's Creation*, (FaithWords, 2017), "Biological vs. Mechanical."

The Healing Sanctuary of the Wild

1. Paul A. Sandifer, Ariana E. Sutton-Grier, and Bethney P. Ward, "Exploring Connections among Nature, Biodiversity, Ecosystem Services, and Human Health and Well-being: Opportunities to Enhance Health and Biodiversity Conservation." *Ecosystem Services* 12, (2015): 1–15, https://doi.org/10.1016/j.ecoser.2014.12.007 .

2. University of East Anglia, "It's Official—Spending Time Outside Is Good for You," *ScienceDaily*, 6 July 2018. www.sciencedaily.com/releases/2018/07/180706102842.htm. Source: Science Daily / University of East Anglia

3. Peter Wohlleben, *The Heartbeat of Trees: Embracing Our Ancient Bond with Forests and Nature*, translated by Jane Billinghurst (Greystone Books, 2021), Chapter 21.

Nature's Rhythms and Design

1. In 2018, a team of scientists used telescopes on the Canary Islands pointed at two ancient quasars to experiment on entangled photon results from two different islands. JR Minkel, "Quantum Spookiness Spans the Canary Islands," *Scientific American*, March 9, 2007, https://www.scientificamerican.com/article/entangled-photons-quantum-spookiness/.

2. David J. Gross, "The Role of Symmetry in Fundamental Physics," *Proceedings of the National Academy of Sciences* 93, no. 25 (1996): 14256–14259, https://doi.org/10.1073/pnas.93.25.14256 . Copyright (1996) National Academy of Sciences, U.S.A. Used with permission.

3. Kevin Hartnett, "A New Kind of Symmetry Shakes Up Physics," *Quanta Magazine*, 2023, https://www.quantamagazine.org/a-new-kind-of-symmetry-shakes-up-physics-20230418/.

4. Alfred Tennyson, "Flower in the Crannied Wall," 1863.

5. Anna Zasimova, "Golden Spiral Template," Vector, Modified. iStock, gm2048334335-563130837.

6. Φ was first documented in 570 BCE by Pythagoras but made almost mythological by Leonardo Fibonacci in the thirteenth century. Johannes Kepler, who figured out how our planets move, also discovered that the Fibonacci sequence always tends to φ.

Symmetry—The Physicist's Key to the Universe

1. Arvin Ash, "Is Symmetry Fundamental to Reality? Gauge Theory has an Answer," YouTube: Wondrium, June 18, 2022, video, 17:46, https://www.youtube.com/watch?v=paQLJKtiAEE.

2. Arvin Ash, "Is Symmetry Fundamental to Reality? Gauge Theory has an Answer," YouTube: Wondrium, June 18, 2022, video, 17:46, https://www.youtube.com/watch?v=paQLJKtiAEE.

3. *Symmetry Magazine*, Accessed 2023, https://www.symmetrymagazine.org/about.

4. Albert Einstein, "Emmy Noether, NY Times obituary," MacTutor History of Mathematics, from *The New York Times* May 4, 1935, https://mathshistory.st-andrews.ac.uk/Obituaries/Noether_Emmy_Einstein/.

5. George Mestral, Velvet type fabric and method of producing same, U.S. Patent US2717437A, filed 15 October 1952, and issued 13 September 1955, https://patents.google.com/patent/US2717437A/en.

6. "Inspired by gecko feet, scientists invent super-adhesive material." 2012. Phys.org. https://phys.org/news/2012-02-gecko-feet-scientists-super-adhesive-material.html.

7. August G. Domel, Mehdi Saadat, James C. Weaver, Hossein Haj-Hariri, Katia Bertoldi, and George V. Lauder, "Shark Skin-Inspired Designs that Improve Aerodynamic Performance," *Journal of the Royal Society* 15. No. 139 (February 2018): 20170828, doi: 10.1098/rsif.2017.0828.

8. As noted on the copyright page, all references to "create" that denote something Created from nothing will be capitalized.

9. Maurice A. Finocchiaro, ed., *The Galileo Affair: A Documentary History*, trans. Maurice A. Finocchiaro (University of California Press, 1989), 146.

10. James Hannam, *God's Philosophers: How the Medieval World Laid the Foundations of Modern Science* (Icon Books, 2009), 303–316.

The Pattern of Uncertainty

1. Note: This is for illustrative purposes only, not investment advice or returns one should expect. But it behooves us to save early and let our money work for us.

2. VvoeVale, "Green Sprig of Fern," Photograph, Modified. iStock, gm478322037-26772246.

3. Pierre-Simon Laplace, *A Philosophical Essay on Probabilities*, (Wiley, 1902), 14.

4. Job 42:2, NIV: "I know that you can do all things; no purpose of yours can be thwarted," written between the seventh and third centuries BCE (dates from *The Book of Job* by John E. Hartley, (Eerdmans, 1988), 18.)

5. Michio Kaku, *Quantum Supremacy: How the Quantum Computer Revolution Will Change Everything* (Knopf Doubleday Publishing Group, 2023).

6. Illustration created using Canva Pro elements.

7. Dan Quinn, CC BY-SA 3.0, https://en.wikipedia.org/wiki/Lorenz_system#/media/File:A_Trajectory_Through_Phase_Space_in_a_Lorenz_Attractor.gif, via Wikimedia Commons 2, November 2013.

Life, Meticulously Fine-Tuned

1. Stephen Hawking, *A Brief History of Time: From the Big Bang to Black Holes*, (Bantam Books, 1988), Chapter 8, Paragraph 20.

The Rhythms of Communication in Music

1. Michael Thaut, Gerald McIntosh, and The Dana Foundation, "How Music Helps to Heal the Injured Brain," *Brainline,* posted March 4, 2011, https://www.brainline.org/article/how-music-helps-heal-injured-brain.

2. Robert Wiedersheim, *The Structure of Man an Index to His Past History*, Second Edition, trans. Bernard H. and M., ed. G. B. Howes (Macmillan and Co., 1893).

3. Stephen C. Meyer, *Signature in the Cell: DNA and the Evidence for Intelligent Design*, (HarperCollins, 2010), Chapters 5, 18, and Epilogue.

The Rhythm of Pair Bonding

1. Patti Callahan, *Once Upon a Wardrobe*, (Harper Muse, 2021), 24.

2. Walter Isaacson, *Steve Jobs*, (Simon & Schuster, 2021), Chapter 2, Paragraph 7.

3. Charles J. Woodbridge, *The Chronicle of Salimbene of Parma: A Thirteenth-Century Christian Synthesis*, (Ph.D. thesis, Duke University, 1945), 116.

4. Stephanie Cacioppo, 2023. *Wired for Love: A Neuroscientist's Journey Through Romance, Loss, and the Essence of Human Connection*. Flatiron Books.

5. Anna Brown, "Americans' Views on Dating and Relationships," *Pew Research Center*, August 20, 2020, https://www.pewresearch.org/social-trends/2020/08/20/nearly -half-of-u-s-adults-say-dating-has-gotten-harder-for-most-peo ple-in-the-last-10-years/.

6. Richard Fry and Kim Parker, "Rising Share of U.S. Adults Are Living Without a Spouse or Partner," *Pew Research Center*, October 5, 2021, https://www.pewresearch.org/social-trends/2021/10/05/rising-s hare-of-u-s-adults-are-living-without-a-spouse-or-partner/.

7. Henri C. Santos, Michael E. W. Varnum, and Igor Grossmann, "Global Increases in Individualism," *Psychological Science* 28, no. 9, July 2017, DOI:10.1177/0956797617700622.

8. Jacob Sweet, "New Surgeon General Advisory Raises Alarm about the Devastating Impact of the Epidemic of Loneliness and Isolation in the United States," HHS.gov, May 3, 2023, https://www.hhs.gov/about/news/2023/05/03/new-surgeon-ge neral-advisory-raises-alarm-about-devastating-impact-epidem ic-loneliness-isolation-united-states.html.

9. Plato, *Theaetetus*, trans. M. J. Levett, revised by Myles Burnyeat, edited by Bernard Williams (Indianapolis: Hackett Publishing, 1992), 152a.

10. Joseph Nolan, "8 Leading Causes Of Divorce," *The Healthy Marriage*, April 1, 2022, https://thehealthymarriage.org/leading-causes-of-divorce/.

11. Michele M. Kroll, "Loneliness has Same Risk as Smoking for Heart Disease," *Harvard Health*, 2016, https://www.health.harvard.edu/staying-healthy/loneliness-has-same-risk-as-smoking-for-heart-disease.

12. Charles Eisenstein also used the comparison of our modern age to building the tower of Babel in *The Ascent of Humanity: Civilization and the Human Sense of Self*, (North Atlantic Books, 2013).

13. Charles Eisenstein, *The Ascent of Humanity: Civilization and the Human Sense of Self*, (North Atlantic Books, 2013), introduction, paragraph 26.

The Narrow View of Specialization

1. Edward O. Wilson, *Consilience: The Unity of Knowledge*, (Knopf Doubleday Publishing Group, 1999).

2. Edward O. Wilson, *Consilience: The Unity of Knowledge*, (Knopf Doubleday Publishing Group, 1999), 56.

3. Edward O. Wilson, *Consilience: The Unity of Knowledge*, (Knopf Doubleday Publishing Group, 1999), 56–57.

4. Julia Belluz, Brad Plumer, and Brian Resnick, "The 7 Biggest Problems Facing Science, According to 270 Scientists," *Vox Media*, September 7, 2016, https://www.vox.com/2016/7/14/12016710/science-challeges-research-funding-peer-review-process.

5. Jerold Mande, "Processed Foods Make Us Sick. It's Time for Government Action," *Harvard Public Health Magazine*, March 1, 2023, https://harvardpublichealth.org/nutrition/processed-foods-make-us-sick-its-time-for-government-action/.

6. Sherry L. Murphy, Kenneth D. Kochanek, Jiaquan Xu, and Elizabeth Arias, "Mortality in the United States, 2020." Centers for Disease Control and Prevention, December 2021, https://www.cdc.gov/nchs/products/databriefs/db427.htm.

7. David E. Newman-Toker, Najlla Nassery, Adam C. Schaffer, et al., "Burden of Serious Harms from Diagnostic Error in the USA," *BMJ Quality & Safety*, 17 July 2023, DOI: 10.1136/bmjqs-2021-014130.

8. "Life Expectancy by Country and in the World (2024)," Worldometer, Accessed February 8, 2025, https://www.worldometers.info/demographics/life-expectancy/#google_vignette.

Miracles—Breaking Patterns

1. Erratic Rock is a wise stop before embarking on the trails in Torres del Paine. https://www.erraticrock.com/

2. David Hume, *An Enquiry Concerning Human Understanding*, (Oxford University Press, 1748, 1999), 46.

3. Clive S. Lewis, *Miracles*, (HarperOne, 2015), 161–162.

The Initial Condition of Life

1. Sean Carroll, *The Big Picture: On the Origins of Life, Meaning, and the Universe Itself*, (Penguin Publishing Group, 2016), Chapter 16, paragraph 15.

2. Sean Carroll, *The Big Picture: On the Origins of Life, Meaning, and the Universe Itself*, (Penguin Publishing Group, 2016), Chapter 16, paragraph 8.

3. In 1929, Edwin Hubble, for whom the Hubble telescope is named, made groundbreaking observations about our universe. By studying the redshift of galaxies, he observed they were moving away from each other, providing evidence the universe is expanding and, therefore, had a beginning.

4. Sean Carroll, *The Big Picture: On the Origins of Life, Meaning, and the Universe Itself*, (Penguin Publishing Group, 2016), Chapter 36, paragraph 15.

The Big Bang Evolutionary Theory

1. String Theory and Inflation of the Universe are currently speculative with no direct empirical data. Sean Carroll, *The Big Picture: On the Origins of Life, Meaning, and the Universe Itself*, (Penguin Publishing Group, 2016), Chapter 36, paragraph 27.

2. CERN stands for Conseil Européen pour la Recherche Nucléaire, which is French for the European Council for Nuclear Research. Today, it is known as the European Organization for Nuclear Research, but it retains the acronym CERN from its original name.

3. George Wald, "The Origin of Life," *Scientific American* 191, August 1954, 46.

4. Michael J. Behe, *Darwin's Black Box: The Biochemical Challenge to Evolution*, (Free Press, 2006), Chapter 6, Section "Sisyphus Would Sympathize," Paragraph 9.

5. William F. Marin and Marek Mentel, "The Origin of Mitochondria," *Nature Education* 3, no. 9, 58, https://www.nature.com/scitable/topicpage/the-origin-of-mitochondria-14232356/.

6. Fred Hoyle and Chandra Wickramasinghe, *Evolution from Space: A Theory of Cosmic Creationism*, (Simon & Schuster, 1984), 148.

7. Fred Hoyle and Chandra Wickramasinghe, *Evolution from Space: A Theory of Cosmic Creationism*, (Simon & Schuster, 1984), 24.

8. Steven Zuryn, "Explainer: What Are Mitochondria and How Did We Come to Have Them?" *The Conversation*, September 21, 2017, https://theconversation.com/explainer-what-are-mitochondria-and-how-did-we-come-to-have-them-83106.

The Intelligent Designer

1. The half-life of carbon-14 is 5,730 years.

2. Heather D. Graven, "Impact of Fossil Fuel Emissions on Atmospheric Radiocarbon and Various Applications of Radiocarbon over this Century," *PNAS* 112, no. 31, (August 4, 2015): 9542–9545, https://www.pnas.org/doi/pdf/10.1073/pnas.1504467112.

3. Maris Fessenden, "Climate Change Might Break Carbon Dating," *Smithsonian Magazine*, (July 27, 2015): https://www.smithsonianmag.com/smart-news/climate-change -might-break-carbon-dating-180956062/ .

4. The Holy Bible details a worldwide flood during the time of Noah, but other ancient texts like the Epic of Gilgamesh and Aboriginal stories also include massive floods. The traditions handed down orally change over time, but they are indicators of experiences. For example, Aboriginal tradition includes the story of an ancestor who built a campfire that exploded. Geological sciences believe there was significant volcanic activity during that time and the exploding campfire was their way of describing those events. The existence of so many ancient texts describing massive floods, even in different ways, indicate something major happened on earth.

The Breath of Life

1. Joan Baptista van Helmont, *Ortus Medicinae* (Amsterdam, 1648).

2. Miller's experiments were questioned, as the elements he used were not the theorized elements available in the Earth's primordial soup. Later, Jeffrey Bada built upon Miller's experiments but used the currently theorized elements of primordial soup. While he produced a brown watery liquid full of amino acids, they were not the building blocks for nucleic acids and nowhere near what is needed to Create mitochondria.

3. James Ferris, prebiotic chemist at Rensselaer Polytechnic Institute, Troy, N.Y., on atmospheric electricity: "You get a fair amount of amino acids, . . . What you don't get are things like building blocks of nucleic acids." Douglas Fox, "Primordial Soup's On: Scientists Repeat Evolution's Most Famous Experiment," *Scientific American*, (March 28, 2007): https://www.scientificamerican.com/article/primordial-soup-ur ey-miller-evolution-experiment-repeated/.

4. This quote is commonly attributed to Carl Sagan, but throughout history, people have expressed the concept in various ways, dating back to ancient inscriptions about the Hittites and Egyptians.

5. For more information about origin of life studies and the theories that explore the origination of DNA, see Signature in the Cell by Stephen C. Meyer or Darwin's Black Box by Michael Behe.

The Creation of Novel Traits

1. Stanley A. Rice, *Encyclopedia of Evolution* (Checkmark Books, 2007), 308.

2. Nicola J. Nadeau and Chris D. Jiggins, "A Golden Age for Evolutionary Genetics? Genomic Studies of Adaptation in Natural Populations," *Trends in Genetics* 26 (2010): 484–492. https://www.sciencedirect.com/science/article/abs/pii/S0168952 51000168X

Darwin and the Galápagos

1. Endemism is a species that is only found in a specific geographic location

Finches, Rapid Evolution, and the Limits of Change

1. Scott Solomon, "What Darwin Didn't Know: The Modern Science of Evolution," *The Great Courses* through https://www.wondrium.com/ , Lecture no. 7, "Rapid Evolution within Species," accessed 2023.

2. Stephen J. Gould, *The Panda's Thumb: More Reflections in Natural History* (Norton, 1980), 127.

3. Fred Hoyle and Chandra Wickramasinghe, *Evolution from Space: A Theory of Cosmic Creationism*, (Simon & Schuster, 1984), 141.

4. Fred Hoyle and Chandra Wickramasinghe, *Evolution from Space: A Theory of Cosmic Creationism*, (Simon & Schuster, 1984), 130.

Objectivity and the Complexity of the Eye

1. The quote "If a lie is told often enough, people will begin to believe it," is often attributed to Hitler, but the true attribution is unknown.

2. Charles Darwin, *On the Origin Of Species* (Melbourne and Toronto Ward Lock, 1911), 154–155. Complete paragraph for full context: "To suppose that the eye, with all its inimitable contrivances for adjusting the focus to different distances, for admitting different amounts of light, and for the correction of spherical and chromatic aberrations, could have been formed by natural selection seems, I freely confess, absurd in the highest possible degree. Yet, reason tells me, that if numerous gradations from a perfect and complex eye to one very imperfect and simple, each grade being useful to its possessor, can be shown to exist; if further, the eye does vary ever so slightly, and the variations be inherited, which is certainly the case; and if any variation or modification in the organ be ever useful to an animal under changing conditions of life, then the difficulty of believing that a perfect and complex eye could be formed by natural selection, through insuperable by our imagination, can hardly be considered real. How a nerve comes to be sensitive to light, hardly concerns us more than how life itself first originated; but I may remark that several facts make me suspect that any sensitive nerve may be rendered sensitive to light, and likewise to those coarser vibrations of the air which produce sound."

3. Simon Baron-Cohen, *The Pattern Seekers: How Autism Drives Human Invention*, (Basic Books, 2023), ebook, Chapter 6.

The Complexity of Life

1. Sean Carroll, *The Big Picture: On the Origins of Life, Meaning, and the Universe Itself*, (Penguin Publishing Group. 2016), Chapter 28.

2. Total entropy is when there is no more disorder to occur. See total heat death.

3. Sean Carroll, *The Big Picture: On the Origins of Life, Meaning, and the Universe Itself*, (Penguin Publishing Group. 2016), Chapter 28, paragraph 17.

4. Heat Death of the Universe: One long-term implication of the Second Law is the idea of the "heat death" of the universe. This is a hypothetical state where all energy has become evenly distributed, and no further work can be done. Entropy would be at its maximum.

5. Geiger, Eric. n.d. "The Story of God." Rooted Network. Accessed February 20, 2025. https://www.experiencerooted.com/. Week 8: A Better Beginning, Group Session video

6. Stephen J. Gould, *The Panda's Thumb: More Reflections in Natural History*, (Norton, 1980), 181.

7. Michael J. Behe, *Darwin's Black Box: The Biochemical Challenge to Evolution*, Tenth Anniversary Edition, (Simon & Schuster, 2006).

8. Richard C. Lewontin, *It Ain't Necessarily So: The Dream of the Human Genome and Other Illusions*, (Granta Books; 2000; first published in the USA by the *New York Review of Books*), 142.

9. For more details on the complexity of the cellular processes, see *Signature of the Cell* by Stephen C. Meyer, chapter 5.

10. Boris Schmidtgall, "Evaluating Models on the Origin of Life," 25 May 2020, https://foclonline.org/content-listing?event=801 , video, 44:46, https://youtu.be/QCFigZ_NIwo?si=s575bHMzSsOMZOB-. Used with permission.

11. Stephen Hawking, "Stephen Hawking: 'Science Makes God Unnecessary,'" *ABC News*, September 6, 2010, https://abcnews.go.com/GMA/stephen-hawking-science-makes-god-unnecessary/story?id=11571150.

12. William Hermanns, *Einstein and the Poet: In Search of the Cosmic Man*, (Branden Press, 2013), 14.

13. George Wald, "The Origin of Life," in *Scientific American Magazine* 191 No. 2 (August 1954): 45–46.

14. Jason Thomas Mraz, "Life Is Wonderful," track 1 on *Mr. A–Z*, Atlantic, 2005, compact disc.

Interpreting the Evidence

1. Eilidh Ramsay, "Half a Century Ago, A Rat Theory Predicts the End of Civilisation," *Sciencepost*, 2019, https://sciencepost.uk/2019/01/a-rat-theory-predicts-the-end-of-civilisation/.

2. Maris Fessenden, "How 1960s Mouse Utopias Led to Grim Predictions for Future of Humanity," *Smithsonian Magazine*, February 26, 2015, https://www.smithsonianmag.com/smart-news/how-mouse-utopias-1960s-led-grim-predictions-humans-180954423/.

3. Edmund Ramsden and Jon Adams, "Escaping the Laboratory: The Rodent Experiments of John B. Calhoun & Their Cultural Influence," *LSE Research Online*, January 2008, https://eprints.lse.ac.uk/22514/1/2308Ramadams.pdf.

Science—Only a Drop of Knowledge

1. James Gleick, *Chaos: Making a New Science* (Cardinal, 1991), Chapter: The Butterfly Effect, Paragraph 33.

2. Werner Heisenberg, *Physics and Philosophy: The Revolution in Modern Science*, (HarperCollins, 1958), 201.

3. Erwin Schrödinger, *"Nature and the Greeks,"* (Cambridge University Press, 1954), 93.

4. "Mattias Desmet about Mass Formation and modern society," interview on the Today Show with Tucker Carlson, September 5, 2022, YouTube, https://www.youtube.com/watch?v=ZltdPfal5x0.Desmet.

Beyond Measure—Faith, Worship, and the Human Experience

1. John Steinbeck, *The Grapes of Wrath*, ed. Robert DeMott, (Penguin Publishing Group, 2006), 114.

2. David Foster Wallace, "This is Water," Transcription of the 2005 Kenyon Commencement Address, May 21, 2005, written and delivered by David Foster Wallace. https://web.ics.purdue.edu/~drkelly/DFWKenyonAddress2005.pdf.

3. Matthew 6:21, NIV, "For where your treasure is, there your heart will be also."

4. Werner Heisenberg, *Encounters with Einstein: And Other Essays on People, Places, and Particles*, (Princeton University Press, 1989), 9.

5. Werner Heisenberg, quoting Kepler in *Across the Frontiers* (Harper & Row, 1974), 175.

6. Walter Isaacson, *Einstein: His Life and Universe*, (Simon & Schuster, 2008), Chapter 7, audiobook.

7. Walter Isaacson, *Einstein: His Life and Universe* (Simon & Schuster, 2008), 95.

8. Isaac Newton, *The Principia Mathematical Principles of Natural Philosophy*, trans. I. Bernard Cohen, Anne Whitman, and Julia Budenz, (University of California Press, 1999), Chapter General Scholium, paragraph 4.

9. Max Planck, *Where is Science Going? With preface by Albert Einstein*, trans. and ed. James Murphy, 1991 Dover Edition, originally published by P. Blakiston Son & Co., 1914, 35.

10. Erwin Schrödinger, *"Nature and the Greeks,"* (Cambridge University Press, 1996), 97.

Dissonant Rhythms—Disconnection and Isolation

1. U.S. Environmental Protection Agency, "Report to Congress on Indoor Air Quality," Volume 2. EPA/400/1-89/001C. Washington, DC.

2. "Abundance of Information Narrows our Collective Attention Span," *EurekAlert!*, April 15, 2019, https://www.eurekalert.org/news-releases/490177. Used with permission.

3. Mattias Desmet, *The Psychology of Totalitarianism*, originally published in Belgium by Pelckmans Publishers, 2022 ed., (Chelsea Green Publishing, 2022), Chapter 5, Paragraph 1.

4. Mattias Desmet, *The Psychology of Totalitarianism*, originally published in Belgium by Pelckmans Publishers, 2022 ed., (Chelsea Green Publishing, 2022), Chapter 3, paragraph 30.

5. I Timothy 6:10 (New International Version).

The Pursuit of a Unifying Theory

1. Stephen Hawking and Leonard Mlodinow, *The Grand Design*, (Random House, 2012), Chapter 17, paragraph 7.

2. Stephen Hawking, *The Theory of Everything*, (Jaico Publishing House, 2006), introduction, paragraph 4.

3. Stephen Hawking, *The Theory of Everything*, (Jaico Publishing House, 2006), introduction, paragraph 4.

4. Sean Carroll, "The God Particle—Course on the God Particle and Higgs Boson," *Wondrium*, accessed January 31, 2024. https://www.wondrium.com/the-higgs-boson-and-beyond.

5. "Forces | Universe – NASA Universe Exploration," NASA Universe, accessed January 31, 2024, https://universe.nasa.gov/universe/forces/.

6. Max Planck. Original quote may vary due to translations from *Das Wesen der Materie* [*The Nature of Matter*], a 1944 speech in Florence, Italy, Archiv zur Geschichte der Max-Planck-Gesellschaft, Abt. Va, Rep. 11 Planck, Nr. 1797.

7. René Thom, *Prédire N'est Pas Expliquer* (To Predict is not to Explain), trans. Roy Lisker, (IHES, Edition, 2010). Original French version published by Editions Eshel (1991). https://www.fermentmagazine.org/Stories/Thom/Predire.pdf. 113.

8. René Thom, *Prédire N'est Pas Expliquer* (To Predict is not to Explain), trans. Roy Lisker, (IHES, Edition, 2010). Original French version published by Editions Eshel (1991). https://www.fermentmagazine.org/Stories/Thom/Predire.pdf. 142.

The Design of Communication

1. Psalm 19:1-2, NIV.

2. An engine may function years after it is separated from its creator, but only faith would believe that such a complex machine developed on its own naturally.

Extravagant Splendor—Life After the Rat Race

1. Felicity Aston, *Alone in Antarctica*, printed and bound by CPI Group, 2013, Chapter 6, second to last paragraph.

2. Philip Ball, *Patterns in Nature: Why the Natural World Looks the Way It Does* (University of Chicago Press, 2016), Introduction, paragraphs 3–4.

3. Maris Fessenden, "The Science Behind Nature's Patterns," *Smithsonian Magazine*, 2016, https://www.smithsonianmag.com/science-nature/science-behind-natures-patterns-180959033/.

4. Richard E. Farson and Ralph Keyes, *Whoever Makes the Most Mistakes Wins: The Paradox of Innovation*, (Free Press, 2002), 75.

The Rhythm of Everything

1. Douglas Adams, *The Hitchhiker's Guide to the Galaxy*, (Random House Worlds, 1997). Sci-Fi in which a computer spends years computing the answer to "the Ultimate Question of Life, the Universe, and Everything." The answer was 42.

2. Ed Yong, *An Immense World: How Animal Senses Reveal the Hidden Realms Around Us*, (Random House, 2023), Introduction, paragraph 33.

3. Ed Yong, *An Immense World: How Animal Senses Reveal the Hidden Realms Around Us*, (Random House, 2023), Introduction, paragraph 33.

4. Walter Isaacson, *Steve Jobs* (Simon & Schuster, 2021), Chapter 32, "Yo-Yo Ma."

5. New Living Translation (Carol Stream, IL: Tyndale House, 2015), Romans 1:20.

6. "Ask and it will be given to you; seek and you will find; knock and the door will be opened to you." Matthew 7:7, NIV

7. Matthew 11:28-30 https://www.biblegateway.com/versions/Message-MSG-Bible/ (MSG) Copyright © 1993, 2002, 2018 by https://www.biblegateway.com/versions/?action=getVersionInfo&vid=65

Glossary:

284

. Suzanne W. Simard, David A. Perry, Melanie D. Jones, *et al.* Net transfer of carbon between ectomycorrhizal tree species in the field. *Nature* 388, 579–582 (1997). https://doi.org/10.1038/41557

Appendix:

1. Amber Angelle, "Immune to HIV: How Do They Do It?" *Live Science*, 2010, https://www.livescience.com/9983-immune-hiv.html.